$6

Guide to
American
Caves

GURNEE GUIDE TO
AMERICAN
CAVES

RUSSELL AND
JEANNE GURNEE

A COMPREHENSIVE GUIDE TO THE CAVES IN THE UNITED STATES OPEN TO THE PUBLIC

Zephyrus Press, Inc. • Teaneck, N.J.

Library of Congress Cataloging in Publication Data

Gurnee, Russell H.
  Gurnee guide to American caves.

  1. Caves — United States — Guide-books.   I. Gurnee,
  Jeanne, joint author.   II. Title.
  GB604.G87   917.3'04'926   79-5042
  ISBN 0-914264-29-X
  ISBN 0-914264-30-3 pbk.

This is a new work, first published in 1980.
Copyright © 1980 by R. H. Gurnee Inc.
First printing, 1980
Published by Zephyrus Press, Inc., 417 Maitland Avenue, Teaneck, NJ 07666
Distributed by Caroline House Publishers, Inc., 2 Ellis Place, Ossining, NY
    10562

Printed in the United States of America by Braun-Brumfield, Inc.

CONTENTS

ACKNOWLEDGMENTS

IN 1966 Howard N. Sloane and Russell H. Gurnee prepared the first comprehensive guide to the show caves of America, entitled *Visiting American Caves*. This was an effort that enlisted Jeanne Gurnee, Lucille Sloane, and numerous members of the National Speleological Society, as well as cave owners all over the country. In the twelve years since the book was published, many changes have taken place. Saddest of these is the passing of both Howard and Lucille Sloane, two good friends, avid cavers and tireless workers. Howard's files, notes, enthusiasm and interest were the driving force behind the book.

Visiting American Caves has long been out of print. It was evident that such a guide to caves was still necessary, but a mere revision of the original text would not be enough, because of the opening of new caves, the closing of others, and the many other alterations brought about by the passage of time. Thus we began again from the beginning, though still feeling Howard Sloane's guiding influence and incorporating much of the background material about cave formation and cave decorations from *Visiting American Caves* into this new book.

We would like to thank Crown Publishers, Inc., who released the rights to *Visiting American Caves* to us; we also wish to thank all those members of the National Speleological Society and National Caves Association who again answered our queries about the status of caves now open to the public. Questionnaires were sent to all owners of private and public show caves; our thanks go to those who took time to send us detailed current information. The photos which appear in this book were also supplied through the courtesy of cave owners and operators. We thank them for permission to use these, and in particular: the National Park Service, for photos on pp. 6, 13 top and bottom, 25, 60, 141, 148, 169, 186; Arkansas Publicity and Parks Commission, for the photo on p. 49; Mississippi Palisades State Park, for the photo on p. 82; Walker, Missouri Commerce, for the photos on pp. 130 and 137 top, Chattanooga Convention and Visitors Bureau, for the photo on p. 200; and Virginia Chamber of Commerce, for the photo on p. 225.

It is not possible to acknowledge all of the sources of information for this book, since the caves mentioned have been the subject of research by thousands of people over the years. Some individuals, however, have offered so much information about a particular area that we would like to extend our special thanks: William Varnedoe and Terry Tarkington, the caves of Alabama; William R. Halliday, the caves of California, Washington, and a dozen other areas; John Thrailkill, Kentucky caves; J. Harlan Bretz, Jerry Vineyard and Dwight Weaver, Missouri caves; Roger Brucker, caves of Ohio; Ralph Stone, caves of Pennsylvania; Gary Soule, who compiled a list of show caves giving information on "formerly commercial" caves as well as leads to additional caves; Robert Armstrong for leads to artificial caves; Barbara Munson for National Caves Association member information; Father Gilbert Stack, caves of South Dakota; Thomas Barr, Roy Davis and many more, the caves of Tennessee; Charles Mohr, Orion Knox and Jack Burch, the caves of Texas; Henry

Douglas, William McGill, and John Holsinger, the caves of Virginia; and William Davies, who supplied information in his remarkable book *Caves of West Virginia.* Special thanks go to Bruce Sloane, who copyedited the manuscript and offered us critical and constructive suggestions.

In the past few months we have traveled over 15,000 miles visiting show caves in the United States that we had not previously seen, and revisiting those we knew visitors would be particularly interested in. We sincerely thank the cave owners, who extended courtesies and friendship to us. The owners, operators, and guides — as well as the caves themselves — left a lasting impression on us. It is an impression we hope we have conveyed to you in the descriptions given in the *Gurnee Guide to American Caves.*

RUSSELL AND JEANNE GURNEE

INTRODUCTION

PEOPLE IN THE UNITED STATES have enjoyed visits to commercial, or show, caves since early in the nineteenth century. Caves in Virginia, Kentucky, and Tennessee were open in 1825, when guides with lanterns, lamps, and torches led groups through various corridors and rooms. At that time trails and pathways were only slightly adapted to the needs of visitors, so a special costume was required by the hardy enthusiasts who crawled, crouched, ducked, and scrambled through the sometimes low and narrow passageways. Among the customary hazards of such a trip were singed hair and clothes covered with candle wax.

Early visitors discovered that a uniform temperature exists within a cave and that this, together with air devoid of pollen and odors, was exhilarating to the senses and produced a feeling of well-being. People seemed capable of exertion that they found impossible outside the cave. The notion that the air within caves possesses some therapeutic value was so strong that it encouraged experiments with patients having respiratory diseases. In Mammoth Cave, Kentucky, such patients were quartered in stone houses as part of a "cure." In the light of present knowledge, it seems impossible that these patients could have benefited from such treatment.

Today, caves open to the public are generally lighted and usually have guide service. Over 200 public caves in the United States provide the opportunity to visit the world below ground. In many instances extensive and costly alterations have been undertaken to make the cave easily accessible, so that within the short span of an hour you can be introduced to an experience unlike anything else in nature. Here under the earth, on graded trails and along guided ways, artificial light reveals the beauty of naturally decorated halls and corridors. Dripping water, sparkling crystals, and water-worn walls present a changing diorama of which you become a part. At each turn a new view is seen, and with each step a different perspective appears before your eyes. The experience of visiting a cave arouses the curiosity and sense of exploration in each of us; it is an adventure of benefit and interest to young and old alike. The purpose of the *Gurnee Guide to American Caves* is to help you select the cave or caves to include in your itinerary and to prepare you to make the most of your visit.

It would not be possible in a small guidebook to give a detailed description of each show cave of interest. Besides, no description can replace the experience of actually seeing, and each person will react differently to what he sees. We have therefore limited ourselves to providing some practical information about each cave, together with a little historical background. With this information you can plan your trip to include whatever seems most convenient and most interesting to you.

If your visit arouses a yen to explore wild caves, we urge you to get in touch with the National Speleological Society, Cave Avenue, Huntsville, AL 35810. The NSS office will let you know which of the Society's grottoes, or chapters, is nearest you. You are welcome at chapter meetings, and you may find that your casual inquiry will develop into a rewarding hobby.

Another organization, the National Caves Association, is composed of show cave owners and operators. It functions as an excellent liaison between the public and show caves. We suggest that those wishing information about caves open to the public contact the NCA office: National Caves Association, Cumberland Caverns, McMinnville, TN 37110.

THE CAVE WILDERNESS

THE INTERIOR OF A CAVE is perhaps one of nature's most fragile environments. It is a world in delicate balance, where the temperature is almost constant, where it has taken many years to create formations from small mineral-laden drops of water, and where the cave's seemingly static condition sustains unusual biological life forms. The intrusion of man can tip this balance, causing massive changes in the cave biota, in the growth of formations, and in the very nature of the cave architecture.

In gathering information for this book, we revisited show caves we had not seen in over twenty years. We looked at them with different eyes — from the viewpoint of our having seen more than 2000 caves — and a number of changes were evident.

Architecturally, the caves look the same. Improvement in lighting technology is evident in many caves, providing a longer view of large rooms, making them appear more dramatic and impressive. Conveniences for the visitor have improved; the trails, ramps, hand rails and surface accommodations are on the average better. Many caves now provide excellent guides who offer interesting and accurate information about the caverns and their formation, for the visitor of today is a more enlightened individual about caves than were his parents.

The picture is not all positive. Many caves have suffered under unknowing management in combination with visitors who do not understand that to remove a formation is to destroy this form of nature's handiwork forever. Trees can be replaced in time by reforestation, but cave formations often never grow again if the water and mineral source is no longer in the same position. The removed formation "dies" and often becomes relegated to a drawer and finally tossed out. In the cave, the formation can be alive, crystalline, growing and an integral part of the total underground landscape.

Show caves of the world have been displayed for only a brief portion of their total existence. Two hundred years is hardly any time at all in the overall geologic picture. Unfortunately many caves once exhibited have not had the protection they deserve and are already "formerly commercialized" — now only a shambles of scattered broken formations.

There are today some enlightened cave owners who understand the fragile nature of the caves in their care and work to preserve them. Luray Caverns, privately owned and exhibited for more than a hundred years, is today probably in as fine a condition as when it was discovered. This required an investment in time and money to clean and protect it. As with any masterpiece, show caves require continual maintenance and attention.

The continued success and preservation of each show cave will depend on the cooperation of the cave owner, as custodian of this fragile landscape, the cave guide who is the sole agent to give an interesting, factual and exciting interpretive message to the visitor, and the visitor himself, who while enjoying the cave landscape leaves it untouched so that his children and children's children may also have the joy of the unique underground experience.

3

HOW TO USE THE GURNEE GUIDE

Every effort has been made to include all caves open to the touring public, both privately owned show caves and those operated by federal, state, county or local agencies. The criterion used for inclusion is whether a family group could get out of its car and enter the cave without needing elaborate equipment or change of clothing. All such caves are included, whether or not a fee is charged for admission.

Information was obtained from the authors' own lists, from cave owners and operators, and from members of the National Speleological Society, as well as from personal visits and questionnaires sent to all caves. Verification was obtained wherever possible of the information given in the returned questionnaires and in brochures furnished by the caves. Every attempt has been made to include all caves meeting the criterion given above, and to be accurate and up-to-date. The authors would appreciate being informed of additions or changes for inclusion in later editions of this guide.

If you would like more detailed information regarding a particular cave, send your request to the mailing address given under the cave name and you will receive a colorful brochure describing facilities available at the cave site.

States This book is organized alphabetically by states. The caves are listed alphabetically under each state.

Cave Names If you prefer to look up a cave by its name, refer to the Index, where cave names are listed alphabetically. For national parks or monuments, if the cave is the main feature of the park the listing appears under the cave name rather than the park name. If the cave is merely one feature among many others, the name of the park or monument appears first.

Road Directions In visiting caves, be sure to follow the road directions rather than the address given under the cave name, which is the mailing address and sometimes just an office in a nearby town where mail is collected.

Admission Admission charges are not given in this guide because entrance fees change frequently, even as this book goes to press. At present the cost to enter a cave is about the same as the cost of a ticket to a motion picture theater (around $3.00 for an adult and generally half price for children). Nearly all caves have special rates for organized groups.

Parking All caves have parking and rest room facilities.

Photography Most caves offer film for sale and welcome photographers. Some caves have special lighting for photographers.

Lighting All caves are lighted unless otherwise noted. In such cases we tell you whether you are required to furnish your own lighting.

Nearby Facilities and Attractions "Nearby facilities" are usually adjacent to or within a few miles of the cave. The term "all facilities" refers to restaurants, snack bars, gift shops, camping, motels, hotels, cabins, and trailer camps, but does not necessarily include swimming, picnicking, nature trails, and so on. "Nearby attractions" are those within a 30-mile radius of the cave.

Closed Caves Readers who do not find a particular cave listed under a state are referred to the Index. The names of show caves that have been closed, caves on which no information was obtainable, those that could not be reached by phone, or caves that may be opened in the future appear in the Index with an asterisk (*). It can be assumed that such caves are not open to the public at this time.

Temperature   Cave temperatures remain remarkably uniform year round. Summer and winter temperatures in caves in the northern part of the country range between 50° F. and 56° F.; in the southern part of the country they range from 56° F. to 69° F. Caves in higher altitudes are usually colder.

Clothing   It is advisable to inquire at the cave ticket office.regarding temperature inside so that a jacket can be worn if necessary.

National Parks or Monuments   Golden Eagle Passports and Golden Age Passports are available for entrance to National Park Service and Forest Service lands. The Golden Eagle Passport, which is not refundable, permits visitation of a list of park areas for a flat fee for the period of a year. The Golden Age Passport provides a 50 percent discount to U.S. residents who are 62 years of age or older. The Golden Eagle Passport may be obtained by mail from National Park Service headquarters in Washington, D.C. or at regional offices. The Golden Age Passport must be obtained in person at the entrance to areas of the National Park System or Forest Service. Proof of age is required.

Types of Caves   An illustrated section at the beginning of the book will acquaint you with the many different types of caves in the United States.

Formations   Most caves contain formations of one kind or another. Descriptions and illustrations of typical cave formations appear in the section on Cave Formations, as well as elementary information on how they were formed. This section will enable visitors to identify most formations that they might encounter in a cave.

Safety Precautions   In the text we have listed as "self-guided" many caves in state and other public areas. A word of caution is advisable to the uninitiated in visiting such caves. They are unlighted, and no guide is available. Many of them require physical stamina, and can be dangerous to those not familiar with caving techniques. In any cave which extends beyond daylight, adequate provisions for light must be made, and spare lights should always be carried. If one becomes lost or if his lights fail in a cave, the best procedure is to remain in one spot. It is advisable that three persons form a minimum party so that one can go for help, if necessary, without leaving another person alone. Hard hats are advisable to prevent head injuries. Sustaining or nutritious snacks such as raisins might also be carried in case of emergency. Heavy high shoes and suitable clothing are also essential. Always notify someone in attendance on the outside before you enter a show cave which is self-guided. These minimum precautions may prevent serious injury or even loss of life.

TYPES OF CAVES

Limestone Caves

Limestone is a sedimentary rock that was originally formed at the bottom of ancient seas. Hundreds of millions of years were needed to accumulate the deposits of marine life, shells, skeletons, and coral and to compress them into stone and thrust them above the sea, creating the huge limestone regions of the world. Formed by this gradual deposition of many ingredients, including billions of tiny sea creatures, limestones have many different hues, textures, and degrees of purity. A common feature of these rocks is their tendency to be dissolved by rainwater, especially after the water has percolated through acid vegetable matter in the soil. This acid condition causes solution of the rock, and in some instances so dissolves it as to cause caverns beneath the surface. Such caverns range from tiny cracks and tubes, which conduct water to lower levels, to gigantic tunnels and chambers. Nearly 95 percent of the world's caves are found in limestone.

Caves in limestone are, geologically speaking, relatively young. The very factors that are needed to make a cave — solution and erosion — are also the factors that destroy it. It is possible to compare some aspects of the "life cycle" of a cave with those of the animal kingdom. Birth — which took place millions of years ago during mountain-building processes — bends, twists, and fractures the

Limestone cave.

6

limestone that is the matrix for the cave. Childhood occurs when the thousands of cracks and crevices carry the aggressive rainwater through the limestone in search of the water table. Some of these fractures capture more water than others, and as soon as a few begin to dominate, the drainage becomes greater and greater. In youth, the cave is a barren tunnel partially or completely filled with water. Maturity is reached when the water is drained away and dripping

Sandstone Cave, New Mexico. Photo by Russell Gurnee.

water from the land above penetrates the ceiling and forms glistening decorations. All is wet and gleaming. Old age is reached when the atmosphere becomes dry and the formations begin to turn to dust and powder. The cover over the cave has become so thin that there is danger of collapse. Finally in death the ceiling of the cave falls, leaving only a sinkhole, gorge, or valley to mark the site.

Each of these stages takes hundreds of thousands of years. All these conditions can be found in the caves of the United States, but most of the caves listed in this book are at the peak of their beauty and appeal.

Sandstone caves

Sandstone is one of the most common sedimentary rocks. It is so easily eroded by weather and wind that the face of sandstone cliffs is often pitted with shallow holes and pockets. Meandering rivers scoop out great hollows and leave huge overhangs in the sharp bends of a gorge. Sometimes these hollows tunnel completely through the wall to form the arches and natural bridges that are so picturesque in the rimrock country of the western United States.

Such shallow caves were very valuable to the early settlers of the West. Indian cultures developed within the shelter of these caves, and protection from

Sea cave, Sea Lion Caves, Oregon.

weather and enemy attack encouraged the building of permanent settlements. Large cliff houses were made of clay and straw, and elaborate systems of ladders and stairways enabled the early Indians to reach the numerous rooms.

Because the arid climate has preserved these Indian remains in remarkable condition, they can be seen today in their original state in some national parks and monuments.

Sea Caves

Water, wind, and sand are powerful grinding agents that will scour and destroy the hardest rock. The coastline of our country has been pounded by surging seas for hundreds of thousands of years. Broad beaches and rolling dunes shift and change with the wind and tide. Towering cliffs hold back the sea, retreating slowly under the relentless hammering of the surf. Water, seeking out the weakest part of a cliff, grinds and chews away at the rock face until it has burrowed in, booming and spraying against the little hollow until it has collapsed the wall. Occasionally a joint in the rock will permit the water to enter deeper and deeper until a tunnel is cut out — a chamber filled with the thundering surge of the waves and the groaning compression of air.

This is the way most sea caves are formed, and the stone might be of any material. Because the action is principally one of abrasion, in time the hardest granite or the softest chalk may be cut away by the action of the waves.

Gypsum Caves

Gypsum is best known to us in a dehydrated form used as the chief ingredient in wallboard for construction in our homes. Gypsum originally occurs as natural stone, and has many economic uses. It is quite common throughout the world. Gypsum is also soluble, and can be cavernous, although it seldom permits the formation of caves of the size found in limestone.

Traces of gypsum appear in limestone caves, usually as secondary formations and wall decorations, and sometimes in the clay of the floor in the form of crystals and needles. Major caves entirely in gypsum, which are quite rare, are usually found in arid regions of the American southwest.

Gypsum rock is easily scratched — sometimes even with a fingernail. Alabaster, a form of gypsum which has been used for thousands of years for statuary and carved utensils, is also found in caves.

Ice Caves

There are many ice caves in the United States, but only a few are in areas that are accessible to the tourist. These caves are usually found at high elevations and in regions where the mean annual temperature is at or below freezing.

Ice caves are actually natural storage areas for the cold air of winter. The

Gypsum cave, Alabaster Caverns, Oklahoma.

rock covering of the cave acts as insulation against the sun and summer air. The duration of the summer in ice-cave areas is not long enough for the temperature to rise above freezing. The result is that each winter and spring more cold air and water replenish the ice perpetually stored within the cave.

Before the days of modern refrigeration, caves with ice were used to provide the luxury of cold drinks on the hottest days of summer. In Italy, noblemen sent runners to the ice caves in the mountains to bring ice for the banquets and dinner tables of Rome.

Some of the lava tubes in the western United States have large quantities of ice that have formed in the freezing air of the cave. Some of these deposits are thousands of years old. Many of the more delicate formations melt in the fall of the year but they are replaced in the spring as the melting water from the surface drips into the freezing air of the cave.

Crystal-clear icicles and prismatic shapes glisten on the walls of some ice caves. Some of the frost crystals are so delicate that the presence of a person in the room will cause them to melt and shatter. Dripping water causes knobby ice stalagmites and slippery, sloping floors. All the beauty of a winter scene is revealed in the blue-and-white galleries of an ice cave.

Ice cave, Crystal Ice Cave, Idaho.

Boulder Cave, Natural Stone Bridge and Caves, New York.

Boulder Caves

Occasionally caves will be formed by the collapse of a cliff face, the debris piling up at the bottom leaving covered rooms and passages between the rocks. Generally these rooms and crevices are quite small and narrow when first formed. In several instances, however, glaciers have carried debris against a cliff face, fracturing off large sections of the wall. With the melting of the ice, the boulders are gently lowered to the base of the cliff. This action leaves strange balanced rocks, broad roof spans, and passages in among the rocks. These caves, usually small, are curious, and of interest mainly to geologists.

Historical Caves

Man's relationship to caves stretches back into the dim past of prehistory. The earliest permanent home was in the shelter of rock overhangs and the entrances of natural caves. The remains and evidences of generations of early people are valuable records of their lives and times. Fortunately, the nature of caves and their protective ability have preserved these remains, which might otherwise have been lost forever.

Many historical caves and archaeological sites in this country are protected in national parks and monuments and state preserves that range from Russell Cave in Alabama, where archaeologists have found evidence of Indian occupation dating back 9000 years, to numerous southern caves that contain remains of Civil War saltpeter mining only 100 years ago.

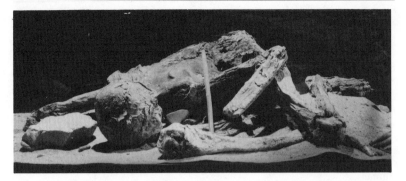

Mummy discovered at Mammoth Cave. Photo by W. Ray Scott.

Lava Tubes

Most caves take hundreds of thousands of years to form, but nature, in a capricious mood, sometimes bypasses this time constraint with the energy and power released by volcanoes. Volcanic eruptions spew forth millions of tons of molten rock in great spasms that can change the face of the earth. These flows of liquid stone surge like rivers down the slopes of the cones and spread out

Lava tube, Lava Beds National Monument, California.

around the base of the volcano like frosting from a cake. The flows occur in a remarkably short period of time, with the great walls of scalding stone sometimes advancing as fast as a man can walk. As the flow spreads out on the level land at the base of the cone, and its rate slackens, the surface of the lava begins to "freeze" and become a solid. Since the core is still hot and liquid, the center continues to flow out the end of this now crudely formed tube. If the conditions are just right for the flow to continue, the snakelike path of this molten river will sometimes extend for several miles. In a very short time the core of the tube will drain out the end, and the entire flow will cool, leaving a tunnel.

In this cooling process, tiny "stalactites" of liquid stone form on the ceiling, almost in an instant, adding decoration to these otherwise barren, subway-like tunnels.

This is the extent of the growth of a lava tube. Born in the fiery flash of the flaming lava, maturing in the moments of the draining of the tube, it remains in a suspended state until the roof collapses or until it is filled again by later lava flows.

Artificial Caves

Artificial caves and grottoes have been built by man since medieval times. Several simulated caves are listed in this guide as being of sufficient interest to merit a visit. There are two categories of artificial caves: those constructed of formations that have been removed from a natural cave, and those constructed of synthetic materials designed to simulate the natural cave landscape.

Artificial cave, the Children's Museum, Indiana.

CAVE FORMATIONS

Stalactites

"Oozing out in drops" is the meaning of this Greek word that describes the icicle-like formations found in caves the world over. Each stalactite begins its life as a single drop of water that forms, swells, quivers, then drops to the floor below. The subsequent drops continue in seemingly endless fashion. If the water is pure and free of minerals, nothing occurs but the patter and splatter of the steady drip. However, if the water becomes saturated with calcium carbonate from the limestone layer above the cave, each drop will leave a tiny deposit of this calcite as it leaves the ceiling. First a tiny ring just the size of the drop is formed; then ring forms upon ring until a slender, drop-sized tube hangs from the ceiling. The formation might continue to grow in this shape until it reaches the floor. These slender "soda straws" sometimes grow to be ten feet long — often as a single crystal. This condition is unusual, however, as ordinarily the tiny hole through the center of the tube will not carry the water from the source above; the water must thus flow down the outside of the straw, adding to its deposit and creating the familiar carrot or icicle shape.

Some stalactites grow to gigantic proportions limited only by the height of the

Stalactites, Ohio Caverns, Ohio.

room and the stability of the joint that supplies it with more calcite for its growth.

The simple ingredients in a drop of water, a tiny bit of mineral matter, the relentless pull of gravity, and centuries of time all work together to create these beautiful natural scenes beneath the earth.

Stalagmites

Every drop of water that falls in a cave must, of course, land somewhere. It can drop in a pool or stream or on a sloping wall, mudbank, or fallen stone. If it falls in water, it ripples and is gone. If it splatters on a steeply sloping wall, it fans out and flows down as part of a sheet of water. As in the development of a stalactite, this action will cause the deposit of calcite; if dripping continues for thousands of years, the wall will be covered with a solid coating called "flowstone."

If the drop falls on a flat, solid surface, the violent action of the water will concentrate the loss of calcite on a spot usually directly beneath its parent stalactite, and a tiny mound will form — the birth of a stalagmite. The rate of flow of water, the distance the drop has to fall, and its impurities are only a few of the factors that determine the characteristics of this budding formation.

Stalagmite, Marvel Cave, Missouri.

A stalagmite is generally rounder and smoother than a stalactite, and has no central tube as its parent does. It may assume a bizarre shape or it may grow to be perfectly straight to a height of ten feet or more. These straight formations are called "broomsticks," and are sometimes no larger in diameter than a silver dollar. If the stalagmite, continuing to build and grow, reaches the stalactite above, the two may join to form a solid pillar of stone reaching from floor to ceiling. This is called a "column." If the ceiling is of great height, and the water splatters over a large area, a stubby, flat, dinner-plate formation resembling a stack of pancakes will be the result.

All of these formations are incredibly old. A tiny nub of stalagmite may have started its growth long before the great pyramids of Egypt were built. A huge column may have begun to form when man was still living in the caves of Europe.

Helictites

Perhaps no other cave formation has proved more puzzling to both visitor and scientist than the bizarre erratics known as "helictites." These strange formations defy gravity, twisting and turning in their growth like tree roots in search of

Helictites, Caverns of Sonora, Texas.

water. There were as many theories as experts among early explorers, the explanations of the strange behavior of these beautiful "stone roots" ranging from wind currents to earth tremors.

A helictite starts its growth as a tiny stalactite—a single drop that leaves a tiny ring of calcite on the wall or ceiling as it drops to the floor. It may continue as a tiny soda straw for many years, straight and true. Then for some unknown reason the chemical composition of the water changes slightly. Some impurities may appear in the water, causing the single crystal structure to change from a cylindrical, prismatic shape to a slightly conical one—each cone crystal fitting into the previous one like an inverted stack of ice-cream cones. This shape is very unstable. As the crystals grow, the direction of the end of the straw might wander from the vertical just as the stack of ice-cream cones might sway or bend. Since this, as well as every hanging formation, has a tiny tube running through it, water will continue to flow through this tube until it emerges as a drop regardless of the direction the tube takes. The result of this wandering can be some of the most curious and often beautiful formations to be found in caves.

There are many questions still to be answered—many examples that seem to prove the experts wrong. No one knows why one little soda straw hanging in a cluster of its fellows will have a corkscrew twist to the end while all the others are straight; or why some huge stalactites will cease to grow as normal carrot-shaped formations, while small helictites sprout out of the sides like fishhooks and rootlets.

Perhaps one day we shall know the trigger that sets off these strange antics of helictites, but that knowledge is not necessary to appreciate the sight of these delicate bits of twisting stone, perhaps the most beautiful of all cave formations.

Rimstone Pools

The crystal-clear water that drips, flows, and forms pools in many caves is actually highly saturated with minerals picked up as the water flows through the limestone rock that covers the cave. These minerals do not dull the brilliance of the pools, but they do trouble those housewives who have to contend with the "hard" water of limestone country. Lime deposits in cooking utensils and "scum" in wash water are caused by this saturation of minerals. There is a bright side, for this very saturation makes possible the magnificent displays of flowstone formations found in caves. Stalactites, stalagmites, columns, shields, and flowstone are merely the residue of this mineral matter transported by water and deposited in bizarre and fanciful shapes.

Usually, when this saturated water flows slowly through a rock-lined channel, its impurities are in equilibrium with its carbon dioxide content. If, however, any disturbance in the stream channel causes the water to ripple or foam, this balance is upset, causing the escape of some gaseous carbon dioxide

Rimstone pools, Massanutten Caverns, Virginia.

(just as bubbles are released when a soda bottle is opened). As a result some calcite is deposited, returning the water to a state of chemical equilibrium. The water flows on as the deposit of calcite continues for perhaps thousands of years until it actually dams up the stream. These dams impound what are known as "rimstone pools," and are perfectly level, controlled in height by the continued growth of the deposit. In the same way that beavers dam up a stream by plugging the holes with clay and sticks, the mineral in the water raises the level of the dam, where the ripples and foam form as the water flows over the top edge.

In some caves, huge dams 100 feet across have impounded thousands of gallons of water; but usually the dams that make rimstone pools are only a few feet across or on sloping surfaces a few inches long. These limpid pools, like terraced fields of rice paddies, gracefully lower the water in ladder-like steps to the continuing stream below.

Fossils

Limestone, formed from the deposit of millions of ancient sea creatures and debris, sometimes captured the remains of these sea animals intact. Many kinds of seashells and coral were entombed in this sedimentary rock, and the outlines of their form have been preserved in solid stone. Many of these remains were destroyed during the formation of the stone, but some of their

Crinoid fossils, Crystal Caverns, Missouri.

forms were replaced by minerals, and thus became harder than the surrounding rock.

Since most caves form in limestone rock, fossils may often be seen projecting from the walls and ceilings. One fossil sometimes seen is the stalk of a sea animal called a crinoid. This animal lived in shallow seas. It sometimes attached itself to the sea floor and had a long stem-like body that reached upward toward the surface of the water. The stalks look like a series of dough-nut-shaped rings joined together in a continuous tube.

Bones of extinct vertebrate animals (much later in geologic time than the primitive sea creatures) are also found in caves. These bones were deposited after the cave formed, and might be the remains of trapped animals that fell into the cave or were dragged into the cave by predators; or the remains might have been washed into the cave by streams or rain.

Gypsum Flowers, Selenite Needles, and Anthodites

Some of the most unusual formations found beneath the earth are also the most beautiful. In certain dry areas of some caves, gypsum will form on the walls and make a crust that sparkles and glistens in the light of the lamp. In some cases tiny patches will seem to sprout out into beautiful, graceful stone flowers that appear to be extruded form the wall. It is not unusual to find twists and curls projecting 10 to 12 inches without any support. These gypsum flowers are extremely fragile, and will sometimes fall of their own weight.

Selenite needles, Cumberland Caverns, Tennessee. Photo by Russell Gurnee.

Anthodite flowers, Skyline Caverns, Virginia.

In the same general areas that support gypsum flowers, crystals are sometimes found that seem to form in the moist air of the cave. Selenite needles apparently form from a seed crystal on the floor and then, seemingly unattached to anything, grow as a single straw sometimes to 2½ feet in length. These needles occur in piles on the floor like jackstraws, completely unattached and loose.

Shield, Grand Caverns, Virginia.

Anthodites are often described as the stone flowers of the caves, and they certainly have the delicate and fragile appearance that would give them that name. Hairlike crystals radiate from a central point, making little blooms of crystals on the wall or ceiling. Anthodites, like most speleothems, are composed of calcium carbonate, but the calcium carbonate is in a different crystal form. This other form is the mineral aragonite. It is the different growth behavior of aragonite that gives the anthodite its unique shape.

Shields

Curiously round and flat formations occasionally found in limestone caves are known as "shields" or "palettes." These disk-shaped layers of calcite project from the walls and ceilings (and occasionally floors) of rooms and passageways. The actual formation of these objects could be shown in a cutaway section revealing two flat disks sandwiched together. Water flows between these two plates—which do not adhere—and at some time during the growth process, stalactites begin to form along the edge. The question of why these disks form, why they do not cement themselves together, and how they can form at different angles to the bedding and joints of the parent limestone has given rise to conflicting answers.

One thing seems sure. They are found only in certain caves and in certain areas. It is possible that either the limestone matrix or impurities in the stone above the cave influence the growth of these beautiful decorations, which sometimes reach ten feet in diameter. It is also possible that minute earth movements and hydrostatic pressure play a part in their growth.

Boxwork

In the formation of mountains and continents, tremendous forces are at work that will bend, twist, and shatter the hardest stone, and fracture the most massive rock. Limestone, forming in huge sheets and layers, is very vulnerable to fracture. Any warping or twisting of the beds sends cracks and joints through the layers, sometimes for many miles. Severe pressures often fracture the stone into small checkerboard-like layers so porous that water will stand in, or flow through, this maze as it would through gravel. Since there is no major conduit for the water to select, all the joints tend to fill uniformly with calcite carried in the water. These cracks eventually become completely filled with veins of calcite. Further growth is not possible once all the cracks are cemented together. The fractured layer then becomes whole again.

If other conditions are then just right for the formation of a cave in that area, the solution of the rock may take place, leaving little ribbons of calcite projecting from the walls and ceiling. Some of these ribbons are so thin that they are translucent, yet they are hard enough to withstand the solution of the limestone matrix that gave them their shape.

Boxwork, Wind Cave National Park, South Dakota.

Cave Pearls

Cave pearls are rare and exotic formations. They are often overlooked, and inadvertently become trodden underfoot in the dim light of a "wild" cave or destroyed by the trail builder in a commercial cave. Disguised as water-worn pebbles in a pool, they sometimes escape the scrutiny of visitor and guide alike. Occasionally they are found in an ideal setting, where they glisten and turn in their "nest" as the water splatters from the ceiling.

Cave pearls, technically called "oolites," are formed by the deposit of calcite (dissolved limestone) on a tiny bit of sand or a pebble. This same material forms stalactites and stalagmites and covers the floor with a pavement of "frozen stone." However, one element is required in the growth of a cave pearl that is missing in the growth of all other cave formations. That element is movement — the motion of the particle that serves as a nucleus for the pearl and causes the deposit of material to form in uniform layers around the "seed."

Most cave pearls form in a shallow pool beneath a steady flow of water. This constantly running waterfall causes the sand and gravel to be suspended by the turbulence of the water. The agitation causes some of the calcite held in the water to deposit on the minute bits of sand, until they are covered with the

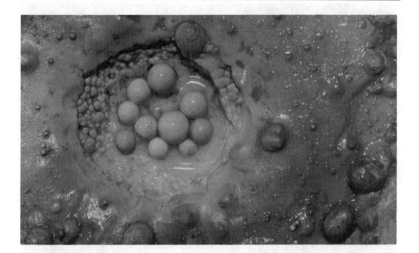

Cave pearls, Carlsbad Caverns, New Mexico.

material and look like tiny beads. After many years of this growth, they sink to the bottom from their own weight. If the pool is shallow enough, the running water will still turn these little spheres so that they continue to grow until they finally get so heavy that they adhere to the bottom and become part of the pool.

During this growth period the pearls are perfectly free and separated from the stone floor. They are suspended on a film of water that permits them to rotate round and round, sometimes forming perfect spheres or even tiny pointed tops that fit into tapered holes in the floor. Most cave pearls are smaller than marbles, but occasionally they grow to the size of grapefruits.

Like most cave formations, these pearls have no value out of their environment because they must remain wet to retain their luster. (They will dry out and turn to powder when taken from the cave.) Their real beauty and rarity are in the place of their birth where the ceaseless splattering of the water and the sparkling drops need only the light of the lamp to reveal their beauty.

Popcorn, Lily Pads, and Dogtooth Spar

Under certain conditions in highly saturated atmospheres within a cave room, curious grapelike clusters not unlike popcorn form on the walls, formations, and even the floor and ceiling. The presence of clear, still pools in these rooms leads to speculation that these speleothems might form beneath the water. It is generally accepted, however, that popcorn forms in the air and that the apparent stratification of the line of growth is due to stratification of the saturated air within the room.

Popcorn, Meramec Caverns, Missouri.

Lily pads, Onondage Cave, Missouri.

Dogtooth spar, Sitting Bull Crystal Caverns, South Dakota.

The waterline of some pools might be emphasized by a thin layer of calcite that forms at right angles to the sides and projects out into the pool. These flat projections sometimes reach large proportions, and are popularly called "lily pads."

Occasionally, in rooms that are completely submerged, the floor, walls, and ceiling will be covered with crystals of calcite. In such rooms, the crystals are free to develop their natural scalenohedral shape. These angular crystals, called "dogtooth spar," sometimes grow to be four or five inches long, and project from the walls and ceilings. They vary in color from white to pale yellow and orange to onyx.

Moon Milk

One of the strangest cave formations is the semiliquid, cheeselike material popularly called "moon milk." This secondary formation is seen on the floors, walls, and ceilings of certain caves. Though similar in many ways to other deposits, it is unique in that it does not harden or turn to stone.

Analysis of this material shows it can be any of several minerals. Theories as to its origin are mixed. It is possible that the formation of this material is not chemical, but rather by bacterial action capable of breaking down the stone to this semiliquid state. The mystery is further compounded by the discovery of moon milk as a dry powder and as a kind of granular snow in certain parts of some caves.

Breakdown

Old age and decay do not attack an entire cave at once. They creep up on it bit by bit as surface valleys cut down into the limestone. Cave passages often end against heaps of fallen limestone blocks at the intersection of the passage with a hillside. These fallen blocks are termed "breakdown."

Breakdown blocks are seen in most caves. They vary in size from pebbles to chunks the size of a house. The piles of breakdown may be only a few scattered fragments or may take the form of breakdown mountains sometimes 100 feet high over which the tourist trail climbs in a series of switchbacks.

There are many processes that trigger breakdown, but most of them occur either very early in the history of the cave or very late. Rockfall is very rare during the mature years of the cave, and it is during the mature years that caves are seen by visitors. Dry cave passages with their naturally arched roofs, and without the influence of man-made stresses from drilling and blasting, are remarkably stable structures. While you may look with awe at the mountain of rubble that fell thousands of years ago when the present landscape was formed, you need not glance apprehensively at the ceiling.

Cave Fills

All the speleothems you will see in caves have been deposited by some sort of chemical reaction after the cavern itself was formed. Less attractive, perhaps, but equally important, are the cave fillings that have deposited by the mechanical action of water. These are the layers of sand, silt, and clay that cover the floors of most caves.

In many commercial caves, trails for the visitors have been prepared by making trenches up to six feet deep in the cave fills. As you walk comfortably through these miniature road cuts (in a passage that the early explorers traversed flat on their stomachs), the fills can be seen in cross section. They are layered like surface rocks, and each layer may be different in color and texture from the ones above and below it. These strata of loose sand and clay record the story of the water that passed through the cave in its youth. Coarse gravels may mean rough, fast-moving streams; fine clay and mud may have been deposited by slowly percolating water with perceptible current. In reading the riddle of the cave's early history, the scientist finds more clues locked in the unattractive fills than in the glistening stalactites.

The fills also serve as containers for minerals and graves for animals. Gypsum crystals grow in the fills of some caves. The crystals of saltpeter, mined in many southern Appalachian caves in the early nineteenth century, occur almost exclusively in the fills. And graves? Animals sometimes fall into caves through pits and sinkholes, or crawl into the caves to die. Skeletons of animals thousands of years old are fairly frequently found buried in the cavern fills.

LANDSCAPE MAP

WE RECOMMEND that you refer to this landscape map when you are looking for caves in a particular region, so that you can see the conditions that have influenced their formation. More detailed geologic and topographic information can be found on the first page of each state's cave listing.

SHOW CAVES OF THE UNITED STATES

EACH CAVE now open to the public is shown on this map of the United States. Look under state headings for individual maps, road directions, route numbers, and descriptions.

ALABAMA

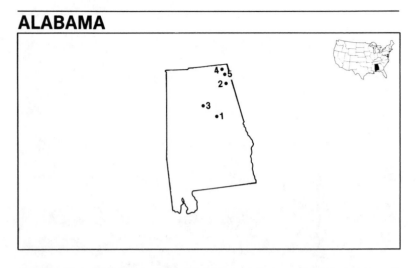

1 DeSoto Caverns 2 Manitou Cave 3 Rickwood Caverns State Park 4 Russell Cave National Monument 5 Sequoyah Caverns

THE NORTHERN PART OF ALABAMA, where the show caves of the state are located, contains folded rocks of the Appalachian Mountains that are more than 300 million years old. Historically the caves of this region have served as shelter for Indians and the oldest known Woodland Indian sites of the eastern United States have been found in these caves. During the War Between the States, many of the caves were mined for saltpeter to supply gunpowder for the Confederate Army.

DESOTO CAVERNS

Mailing address: Route 1, Box 50A, Childersburg, AL 35044 *Phone:* (205) 378-7252   *Directions:* 5 miles east of Childersburg on State Highway 76.

In 1540 Don Hernando de Soto made the first expedition into the "roaring wilderness" of the North American continent. Riding the first horses ever seen by the astonished natives, he paused in his journey westward in the wooded hills of southern Alabama. It is conjectured that he visited a cave known to the early Indians as KyMulga Cave and this established the legend that the cave visited was that now called DeSoto Caverns. Today a paved road leads to the entrance and a small museum displays Indian artifacts taken from the floor of the cave and vicinity of the entrance. The large natural entrance and the impressive entrance chamber were well known to the trappers and explorers who followed De Soto. Several signatures dating from the 1700s are still visible on the walls and formations. The present management is improving the trails for the convenience of visitors.

¶OPEN: February to March — weekends only; April to September — every day

DeSoto Caverns, Alabama.

¶GUIDED TOUR: 1 hour   ¶ON PREMISES: camping and picnicking   ¶NEARBY: in Childersburg, most facilities   ¶NEARBY ATTRACTIONS: Andrew Jackson's Fort Lashley, Logan and Martin Lakes, Cheaha State Park (highest point in Alabama), Lay Dam, Coosa (De Soto's Camp), and marble quarries.

MANITOU CAVE

Mailing address: 405 Alabama Avenue N, Ft. Payne, AL 35967   *Phone:* (205) 845-1332   *Directions:* Fort Payne, off U.S. 11 within city limits of Ft. Payne and one mile from Interstate 59.

Manitou Cave was discovered by Cherokee Indians and was believed to house the "Great Spirit" of the Cherokee Nation. Opened in 1878, this cave is one of the earliest tourist attractions in Alabama, and the photograph shows the entrance at that time. During the War Between the States, saltpeter was mined in the cave to make gunpowder for the Confederate Army; in the 1890s a railroad spur line brought visitors to the area, which became popular as a resort. Today visitors walk over steel bridges spanning the small stream that flows through the cave. A dam constructed in 1899 impounds Rainbow Lake. The guided tour includes a Cherokee Ceremonial Room whose walls are stained by the fires of Indians and which also contains original Indian writings.

Manitou Cave, Alabama.

¶OPEN: June to September   ¶GUIDED TOUR: 1 hour   ¶ON PREMISES: snack bar, gift shop, picnicking. The cave is six blocks from all other city facilities. ¶NEARBY ATTRACTIONS: DeSoto Canyon area, Sequoyah Caverns, Buck's Pocket State Park, DeSoto Falls, Little River Canyon, Guntersville Lake, Russell Cave National Monument, Noccalula Falls.

RICKWOOD CAVERNS STATE PARK

Mailing address: Route 3, Box 340, Warrior, AL 35180   *Phone:* (205) 647-9692
Directions: Hayden Corner Exit on Interstate 65 about 6 miles north of Warrior.

A scenic drive off Interstate 65 leads to a heavily wooded park with outcroppings of limestone blocks and pavements characteristic of cave country. In Rickwood Caverns, developed in 1930 as a private venture, the visitor covers more than a mile of passageway. It is a winding, twisting tour, the first part of which leads deep beneath the mountain. The solution-formed walls are dry and clean. Even in the larger rooms there are few formations, but as the trail turns away from the mountain and descends beneath the valley floor it becomes wetter, and active formations are encountered. The last half of the tour descends to a clear pool which is the water supply for the park above. Exit is made up a spiral staircase that returns to the Visitors' Center and gift shop. There are plans to install an elevator to avoid the 100 steps in the cave, but if taken gradually the ascent is not difficult. The cave was purchased by the State in 1974, and the surface facilities have been improved for the enjoyment of summer visitors who wish to swim (in water from the cave) or picnic in the very attractive wooded setting.

¶OPEN: March to December   ¶GUIDED TOUR: 45 minutes   ¶ON PREMISES: snack bar, gift shop, trailer camp, camping, picnic area, swimming pool. ¶NEARBY: all facilities in Birmingham   ¶NEARBY ATTRACTIONS: Ave Maria Grotto, Russell Cave National Monument, Birmingham, DeSoto Caverns, Natural Bridge.

RUSSELL CAVE NATIONAL MONUMENT

Mailing address: Bridgeport, AL 35740   *Phone:* (1-205) 495-2672   *Directions:* U.S. Highway 72 to County Road 91, then north to Mount Carmel, then right on County Road 75 to entrance.

One of the most significant archaeological sites in the eastern United States is this 300-acre preserve donated to the Park Service by the National Geographic Society in 1961. This area was occupied by Indians 8000 years ago, and evidence of their presence has been found by the excavation of the floor of Russell Cave. Only the entrance area can be visited without special written permission of the Superintendent; but this broad, sheltered opening has provided protection, water and campsite for woodland hunters for thousands of years. A descriptive sound and light show in one of the excavations shows the periods of occupancy in the strata of the dig, and the Visitors' Center contains a display of artifacts and relics taken from the cave. During the summer there are nature walks, demonstrations of fire making and arrowhead manufacture, and also a spearthrowing area for modern hunters to test their ability. The cave is extensive but shows little evidence that early man penetrated far beyond the entrance.

Russell Cave National Monument, Alabama.

¶OPEN: all year   ¶SELF-GUIDED: tour to archaeological site only; no lights required   ¶ON PREMISES: picnicking and hiking trail   ¶NEARBY: restaurant, snack bar, motels   ¶NEARBY ATTRACTIONS: Widow's Creek Steam Plant, Chickamauqua-Chattanooga Battlefield National Military Park, Sequoyah Caverns, Manitou Cave, Chattanooga, Lookout Mountain, Ruby Falls, Rock City.

SEQUOYAH CAVERNS

Mailing address: Valley Head, AL 35989   *Phone:* (1-205) 635-2311   *Directions:* Off U.S. 11 and Interstate 59 about 35 miles south of Chattanooga, Tennessee.

This historic cave, named for the famous Cherokee Indian who invented the only written alphabet of the Cherokee language, was locally known as Ellis Cave. It enjoyed a moderate reputation as a site for Sunday afternoon visits — until the coming of Clark Byers, present owner and developer. In 1965, after 34 years of painting Rock City signs on barn roofs all over the southern United States, Byers turned his talents to the development of the cave. His imagination and creativity, in combination with the abilities of his partner, produced an effect that is unusual among the developed caves of the country and has greatly increased visitation. The stalactites and stalagmites are dramatically

Sequoyah Caverns, Alabama.

reflected in pools of water throughout the cave, and the skillful lighting of the rooms creates an illusion of depth. The cave's natural pools have black, mirror-like surfaces. The clear reflections of objects above make waterfalls appear to flow upward and ceiling crevices look like deep canyons.

¶OPEN: all year from 8:00 A.M. to 7:00 P.M. in summer, and from 8:30 A.M. to 5:00 P.M. in winter   ¶GUIDED TOUR: 1 hour   ¶ON PREMISES: gift shop, camping, picnicking, lounge   ¶NEARBY: restaurants, snack bar, motels, trailer camp   ¶NEARBY ATTRACTIONS: Manitou Cave, DeSoto Falls, Lookout Mountain, Ruby Falls Cave, Russell Cave National Monument.

ALASKA

THERE ARE NO SHOW CAVES in Alaska. The Panhandle section has a few small caves and the Arctic Slope to the north has several sea caves formed by the waves of the Arctic Ocean.

ARIZONA

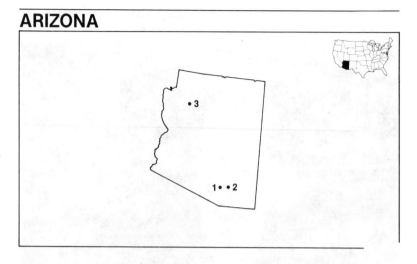

1 Arizona-Sonora Desert museum (artificial cave) 2 Colossal Cave 3 Grand Canyon Caverns

THE MAJOR FEATURES OF Arizona's topography were established in relatively recent geologic time — between 70 million and 130 million years ago. Much of the area was covered by a shallow sea, and the sedimentary rocks show fossils including sharks' teeth and sea shells. Perhaps the world's most spectacular example of erosion is the Grand Canyon of northern Arizona where the Colorado River and the Little Colorado River join to form the 200-mile-long canyon.

About 10,000 years ago climatic changes caused most of the state to become relatively arid. Indians occupied shelter caves eroded out of the soft sandstone and built elaborate cliff dwellings sometime after 1100 A.D. The few limestone caves open to the public are generally dry and typical of most of the caverns found in the state.

ARIZONA-SONORA DESERT MUSEUM (artificial cave)

Mailing address: Route 9, Box 900, Tucson, AR 85704   *Phone:* (602) 883-1380
Directions: About 14 miles west of Tucson on Kinney Road in Tucson Mountain Park.

Since 1952 this privately financed museum has presented exhibits in botany and zoology where the visitor feels that he is a part of the environment. In 1973 an Earth Science Center was added to the facilities and an underground artificial cave constructed. The site appears natural on the surface, and below the visitor has an opportunity to glimpse the cave world. The entrance and exit to the man-made cave are in a natural-appearing wash (dry canyon) which is landscaped with local plants. The tour includes a visit to a wet cave and to a dry cave. There is a crawlway for the energetic; a diorama of mining, rocks and minerals; and displays of earth-building processes — seen from the bottom. Some of the cave

Colossal Cave, Arizona.

formations are made of a special cement and some of fiberglass; much of the breakdown in the cave consists of actual limestone slabs moved to the site. In spite of the fact that the cave is a geological impossibility, a very entertaining transition between displays makes the effect realistic.

¶OPEN: all year 8:30 A.M. to sundown   ¶SELF-GUIDED TOURS: Admission to the

Museum includes the tour of the cave. Tour of the museum — allow 2 hours. Cave tour only — 30 to 45 minutes. Caves are electrically lighted. ¶ON PREMISES: snack bar, gift shop. ¶NEARBY: all facilities in Tucson ¶NEARBY ATTRACTIONS: "Old Tucson" movie location, Colossal Cave, Saguaro National Monument East and West, Picture Rocks, San Xavier Del Bac Mission.

COLOSSAL CAVE

Mailing address: Vail, AR 85641 *Phone:* Colossal Cave No. 1 *Directions:* Interstate 10 about 22 miles east of Tucson.

A few miles south of Saguaro National Monument and partway up the slopes of the Rincon Mountains is the spectacular setting for the entrance of Colossal Cave. Once inhabited by wild animals, visited by local Indians, and used as a hideout by outlaws, this cave is now open to the public. The cave has engendered tales of hidden bank loot and treasure, lost passages that were reputedly closed by trail-building in the 1930s, and a persistent report of 39 miles of passageway. The tour provides enough underground scenery to satisfy the curious; and although the massive formations are now dry, they are imposing to view. The water that dissolved the limestone when the cavern was developing has disappeared, possibly to some lower undiscovered passage. Perhaps this is where the treasure lies hidden and waiting to be discovered.

¶OPEN: all year ¶GUIDED TOUR: 50 minutes ¶ON PREMISES: snack bar, gift shop, camping, trailer camp, picnicking ¶NEARBY: all facilities ¶NEARBY ATTRACTIONS: San Xavier Mission, Tucson National Park, Arizona-Sonora Desert Museum, Old Tucson.

GRAND CANYON CAVERNS

Mailing address: Box 108, Peach Springs, AR 86434 *Phone:* Grand Canyon Caverns #1 through Prescott, Arizona Operator *Directions:* On U.S 66, about 24 miles West of Seligman.

Grand Canyon Caverns (no connection with the National Park) is surprising, for the barren plateau under which it lies hides a subterranean treat. The original entrance, a deep pit-like crevice, was a natural trap for unwary animals. The bones of ground sloths that roamed the region 20,000 years ago have been found at the bottom. Bobcat remains have been found more recently, mummified by the dry cave air. The dryness is unusual, caused by the lowering of the water table by the Colorado River at the bottom of the Grand Canyon only 15 miles to the north. The present water table is 1400 feet down. The cave is the product, however, of the water that once filled the cavity, causing an outstanding display of pure white formations — called snowballs — and golden colored walls. An elevator makes access to the cave pleasant, whereas previous visitors had the adventure of entering with candles via a swinging-stair bridge to the 200-foot level.

Grand Canyon Caverns, Arizona.

Development of facilities next to the cave on Highway 66 provides an oasis for the traveler, including a motel, restaurant, laundry, and airplane landing strip.

¶OPEN: all year, summer 7:00 A.M. to 7:00 P.M., winter 8:00 A.M. to 6:00 P.M. ¶GUIDED TOUR: 45 minutes   ¶ON PREMISES: restaurant, gift shop, camping, motel, trailer camp, picnicking, air strip, post office, and laundry   ¶NEARBY: other facilities   ¶NEARBY ATTRACTIONS: Havasupai Canyon Tour, Kingman Aviation flight over Grand Canyon, Hulapai Indian Reservation, Mead National Recreation area.

ARKANSAS

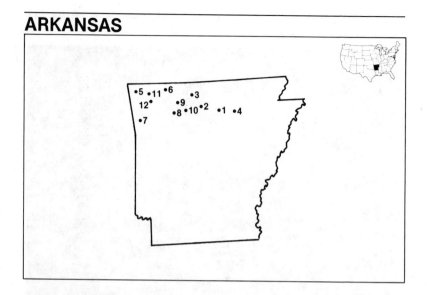

1 Blanchard Springs Caverns 2 Buffalo National River 3 Bull Shoals Caverns 4 Cave City Cave 5 Civil War Cave 6 Cosmic Cavern 7 Devil's Den State Park 8 Diamond Cave 9 Dogpatch Caverns 10 Hurricane River Cave 11 Onyx Cave 12 War Eagle Cavern

ARKANSAS IS DIVIDED almost equally into lowlands and highlands. The lowlands of the Gulf Coastal Plain have no caves, whereas the highlands, part of the Ozark Mountains, have some of the finest caves in the United States. The Ozark Mountain region is underlain by limestones and dolomites with some shales and sandstones. Most of the limestones have been gently folded. Many of the caves are quite large and lavishly decorated with secondary formations. Cave entrances were used for shelter by both man and animals and many show evidence of Indian visitation. Because of the heavy timber growth over the highland area, many previously known caves were "lost" only to be rediscovered by recent cave explorers.

BLANCHARD SPRINGS CAVERNS

Mailing address: P.O. Box 1, Mountain View, AR 72560 *Phone:* (501) 757-2211
Directions: 15 miles northwest of Mountain View, off State Highway 14.

Perhaps no cave in the United States has been as elaborately developed as this fine cavern in the Ozark hills of northern Arkansas. The U.S. Forest Service has spent ten years preparing trails, lights, elevators and surface facilities in order to afford the visitor one of the finest cave experiences in the country. Known for many years as Half Mile Cave, it contains a passage showing evidence of Indian occupation more than 1700 years ago. The stream which flows through the cave appears at Blanchard Springs, where it ran a grist mill in the late 1800s. In the early 1960s two local spelunkers, Hail Bryant and Hugh Shell, discovered the large upper rooms of the cave; as a result the U.S. Forest Service in 1963 began to plan development as part of its recreational program. Today there are two tours offered: The Dripstone Trail provides a bus ride to the lower artificial entrance followed by a walk to the Coral Room and then through the Cathedral Room, and exits up a 216-foot elevator to the Visitors' Center. This tour is the less strenuous of the two trips and shows one of the most dramatic cave rooms on the continent. The Discovery Trail, opened in July of 1977, starts with a descent by elevator, continuing with a walk through part of the Cathedral Room and down a trail to the river passage. This tour passes the natural entrance, views rooms discovered by the Indians, and passes artifacts left by these early explorers. Exit is made through an artificial tunnel where a bus returns the group to the Visitors' Center.

Blanchard Springs Caverns, Arkansas.

¶OPEN: all year except Thanksgiving, Christmas and New Year's Day; May through September 8:00 A.M. to 6:30 P.M., September through May 9:00 A.M. to 4:30 P.M.   ¶GUIDED TOURS: summer—Dripstone Trail, 1 hour or Discovery Trail, 1 hour 40 minutes; winter—Dripstone Trail trip only, 1½ hours. Special group tours are also available by advance reservation—call (501) 757-2213. ¶ON PREMISES: There are no food services or motels in the park, but campsites and trailer hookups are available, as well as picnicking, swimming, fishing, hiking, and an exhibition hall and film in the Visitors' Center.   ¶NEARBY ATTRACTIONS: Bull Shoals Lake, Ozark Folk Center, Buffalo River State Park, Hurricane River Cave, Bull Shoals Caverns, Mountain View Music Festival, Diamond Cave.

BUFFALO NATIONAL RIVER

Mailing address: P.O. Box 1173, Harrison, AR 72601   *Phone:* (501) 741-5443 *Directions:* On State Highway 14 about 17 miles south of Yellville at Buffalo Point.

In 1972 an act of Congress made a 150-mile stretch of the Buffalo River in northern Arkansas a National Park "preserving as a free-flowing stream an important segment of the Buffalo River in Arkansas for the benefit and enjoyment of present and future generations." One of the few sizable free-flowing rivers remaining in mid-America, it looks today much as it did when the Indians roamed the land. Much of the hill country around it is limestone and the river

Bat Cave, Buffalo River State Park, Arkansas. Photo by J. H. Schermerhorn.

has cut a sinuous course, sometimes leaving 500-foot-high walls, as it travels east to join the White River.

At Buffalo Point, the only developed area on the river, there are dining facilities and cabins for rent. At this point, also, access is made to the river for float trips and swimming as well as to camping and picnicking sites. Hiking trails lead to several interesting caves of historical interest: Indian Rockhouse Cave, a one-mile hike from the Park office, and Bat Cave, a quarter-mile farther. These caves are undeveloped and require that visitors use flashlights and wear hiking boots.

There are many other caves in Buffalo National River Park. Information regarding them can be obtained by writing to the Superintendent, Buffalo National River Headquarters, Federal Building, Walnut and Erie Streets, Harrison, AR 72601.

¶OPEN: all year ¶SELF-GUIDED TOUR: about 1 hour; flashlights or other lighting sources required ¶ON PREMISES: restaurant, gift shop, camping, trailer camp, picnicking, float trips on the Buffalo River, swimming, fishing, cabins ¶NEARBY: motels and other facilities ¶NEARBY ATTRACTIONS: Bull Shoals Caverns, Hurricane River Cave, Diamond Cave, Blanchard Springs Caverns, Ozark Folk Center.

BULL SHOALS CAVERNS

Mailing address: P.O. Box 151, Bull Shoals, AR 72619 *Phone:* (501) 445-4300 *Directions:* On State Highway 178, just 4 blocks from Bull Shoals post office. (The town of Bull Shoals is not shown on some state maps. The cave is on Arkansas 178, which runs north from U.S. 62 at Flippin.)

The ticket office of Bull Shoals Caverns is an 1890 train station complete with a steam engine which appears to be just departing. In addition to this, the surface features include a restored mountain village. The caverns are reached by a short trail that ends in a steep grade to the cave entrance. The tour itself is basically on one level. The cave has a history of visitation by Indians, as well as by early pioneers who used it for shelter and as a water supply. There is an underground trout stream and a small waterfall at the end of the tour.

¶OPEN: March 1 to November 1, 8:00 A.M. to 5:00 P.M. ¶GUIDED TOUR: 45 minutes ¶ON PREMISES: gift shop ¶NEARBY: all facilities ¶NEARBY ATTRACTIONS: Hurricane River Cave; Blanchard Springs Caverns; Bull Shoals State Park, Lake and Dam; resort areas; Top of the Ozarks Tower.

CAVE CITY CAVE

Mailing address: Cave City, AR 72521 *Phone:* none *Directions:* On U.S. Route 167 in Cave City.

This small sandstone cave runs directly beneath U.S. Route 167, and the entrance is surrounded by a motel. The passage extends about 300 feet to an underground lake now used as the water supply for Cave City. Caves in sandstone are relatively rare. They do not contain the decorative formations that occur in limestone caverns and so are more of geologic than aesthetic interest.

¶OPEN: all year   ¶SELF-GUIDED TOUR: lights required   ¶ON PREMISES: motel ¶NEARBY: most facilities ¶NEARBY ATTRACTIONS: Blanchard Springs Caverns, White River, fishing.

CIVIL WAR CAVE

Mailing address: Bentonville, AR 72712   *Phone:* (501) 795-2277   *Directions:* 4 miles west of Bentonville on Arkansas 72.

During the Civil War, this cave was used as a supply station for Confederate Army troops under General Van Dorn and provided drinking water from its underground stream. Access to the cave is down a stairway and passageway leading to the stream passage, which is notable for the large rimstone dams that span its width. The present water supply for the surface facilities comes from the cave. The hydraulic ram pump that provides the pressure is visible on the tour.

¶OPEN: all year   ¶GUIDED TOUR: 30 minutes   ¶ON PREMISES: gift shop, camping, picnicking ¶NEARBY: all facilities ¶NEARBY ATTRACTIONS: Onyx Cave, Cosmic Cave, War Eagle Cave, Eureka Springs, Lake Wedington Recreation Area, Pea Ridge National Military Park, Beaver Lake, Withrow Springs State Park, Sulphur Springs.

COSMIC CAVERN

Mailing address: Rt. 4, Box 168, Berryville, AR 72616   *Phone:* (501) 749-2298 *Directions:* Arkansas Highway 21 about 7 miles north of Berryville.

Formerly called Mystery Cave, Cosmic Cavern features a large underground lake stocked with Rainbow Trout. Originally developed in 1925, it was quarried for onyx to be used for jewelry. Fortunately, the depredation of this material has been slight and many fine formations are exhibited. A waterfall room was opened to the public in 1977 and at this point the guide will take family photos with owners' cameras as a memento of the trip.

¶OPEN: all year, 8:00 A.M. to 7:00 P.M. SUMMER; 9:00 A.M. to 6:00 P.M. fall, spring and winter. Closed Thursdays and Fridays in winter and Fridays in spring ¶GUIDED TOUR: 45 minutes. Special night cave tours are available for groups in summer.¶ON PREMISES: gift shop, snack bar, picnicking ¶NEARBY: restaurant, camping, motels, cabins, trailer camp ¶NEARBY ATTRACTIONS: Onyx Cave, Silver Dollar City (Missouri), Eureka Springs' Great Passion

Play, Saunders Museum, Quigley Castle, Table Rock Lake, Beaver Lake, War Eagle Cave, Diamond Cave, Bull Shoals Caverns, Buffalo National River.

DEVIL'S DEN STATE PARK

Mailing address: Rt. 1, West Fork, AR 72774   *Phone:* (501) 846-3716   *Directions:* U.S. 71 to Winslow, then Interstate 74 to West Fork.

This park features a number of sandstone caves, fissures, and crevices. Lee Creek flows through the middle of the park along the valley floor; cabins, a swimming pool, a pavilion, and picnic and camping areas are located along the hillsides. There are no conducted tours to the caves, trips being self-guided, but a trail map is available which shows cave locations. The area is noted for its trails and is the only state park in Arkansas that has an overnight hiking trail. The photograph shows visitors descending into "Devil's Ice Box."

¶OPEN: all year   ¶SELF-GUIDED TOURS: Visitors should bring their own lights.   ¶ON PREMISES: restaurant, snack bar, gift shop, camping, cabins, trailer camp, picnicking, bathing, fishing. Restaurant open May 1 to Labor Day   ¶NEARBY: nearest complete facilities in Fayetteville (20 miles)   ¶NEARBY ATTRACTIONS: Lake Fort Smith, Prairie Grove Battlefield Park, Lake Wedington Recreation Area, White Rock Shores Lake.

Devil's Den State Park, Arkansas.

DIAMOND CAVE

Mailing address: Jasper, AR 72641   *Phone:* (501) 446-2636   *Directions:* Off
State Highway 7 about 4 miles west of Jasper.

A new paved road leads to this cave which has been visited by local people for
more than a hundred years. It was discovered by "Uncle Sammy" and "Andy
Hudson," who chased a bear into the entrance in the early 1800s. Their dogs
followed and in the ensuing battle the dogs and bear were killed. The discovery
of the cave disclosed several miles of passage that were first shown with torches,
then lanterns, and now electric lights. The stream passage has fine examples of
"lily pads" extending out into the center of pools, but for the first part of the
tour there are a great many vandalized formations and blackened walls, caused
by the years of visitation with torches. The cave contains a long, sinuous stream
passage with several places where the visitor must crouch down and "duck
walk" beneath the 3-foot-high ceiling. The end of the tour leads to a series of
well-decorated rooms, clean and unvandalized, with some superior "broom-
stick" formations two inches in diameter and 15 to 18 feet high. The tour lasts for
two hours and requires some strenuous exertion. The last part of the trip justifies
the effort, however.

¶OPEN: all year 8:00 A.M. to 6:00 P.M.; every 30 minutes in summer   ¶GUIDED
TOUR: 2 hours   ¶ON PREMISES: snack bar, gift shop, picnicking, trailer camp,
camping   ¶NEARBY: all facilities in Jasper   ¶NEARBY ATTRACTIONS: Dogpatch
Caverns, Hurricane River Cave, Cosmic Cavern, Onyx Cave, Buffalo National
River, Blanchard Springs, Bull Shoals Caverns.

DOGPATCH CAVERNS

Mailing address: Marble Falls, AR 72648   *Phone:* (501) 743-1111   *Directions:*
State Highway 7, about 7 miles south of Harrison, just north of Dogpatch,
USA.

When this cave was discovered in 1829, the area was isolated and spotted only
with a few settlers' cabins and Indian trails. Now State Highway 7 runs only a
few feet from the entrance, and thousands of tourists visit Dogpatch USA, a
summer attraction, only 1000 feet away across the highway. The cave, developed
in 1950 as Mystic Caverns, featured a trout farm as an additional attraction.
Today the attraction featuring Al Capp's comic strip character Li'l Abner has
transformed this bucolic region into a zany recreation area that expands into
life-sized realism the pictures that Capp drew. Now associated with the
sophisticated Marble Falls Resort, the cave is open as a separate attraction and
provides a very conventional view of a fine cave scene, representative of the caves
of the Ozarks.

    When the cave was being developed and lighted by Jim Schermerhorn in the
mid-1960s, he had completed the wiring and was preparing to install a parking

Dogpatch Caverns, Arkansas. Photo by J. H. Schermerhorn.

lot for visitors when his bulldozer opened another room not connected to the known cave in any way, but only a few hundred feet away. This fine room is scheduled to be opened in the near future and will provide an additional perspective to the tour.

¶OPEN: Memorial Day to Labor Day, 9:00 A.M. to 6:00 P.M. Open weekends during the rest of the year   ¶ON PREMISES: gift shop, picnicking   ¶NEARBY: all facilities   ¶NEARBY ATTRACTIONS: Buffalo National River, Diamond Cave, Hurricane River Cave, Cosmic Cavern, Onyx Cave, Bull Shoals Caverns, Blanchard Springs Caverns, Ozark Folk Center, Eureka Springs resort area.

HURRICANE RIVER CAVE

Mailing address: P.O. Box 93, Western Grove, AR 72685 *Phone:* (501) 429-6200 *Directions:* 16 miles south of Harrison on U.S. Highway 65.

Flowing water has formed this cave, and the present stream emerging from the entrance is the source of the Hurricane Branch of the Buffalo River. Located at the base of a huge limestone cliff, the entrance served as a shelter for Indians and once as a shelter for early settlers during heavy storms. The tour traverses the nearly level stream bed where the scalloped walls, cut by the flowing waters, can be clearly seen. The cave also contains large rooms which have not been subjected to stream action and which are well decorated. The constant water supply from the cave was used in a zinc-mining operation that flourished in the valley during the 1920s. The village of 200 people who were employed in that enterprise has disappeared, but the water of the cave continues unabated.

¶OPEN: April through October, 8:00 A.M. to 6:00 P.M. (Calling beforehand is advisable, as hours fluctuate.) ¶GUIDED TOUR: 45 minutes ¶ON PREMISES: gift shop, picnicking ¶NEARBY: restaurant and all facilities ¶NEARBY ATTRACTIONS: Buffalo National River, Diamond Caverns, Dogpatch Caverns, Bull Shoals Caverns, Blanchard Springs Caverns.

ONYX CAVE

Mailing address: Rt. 1, Box 330, Eureka Springs, AR 72632 *Phone:* (501) 253-9321 *Directions:* U.S. 62 east from Eureka Springs to Onyx Cave sign, then left 3½ miles.

Eureka Springs, once a spa and resort for the infirm who came to take "the waters," now is a resort for vacationing families who like the mountain scenery and clear air of the region. One of the attractions that has continued to appeal to visitors is Onyx Cave, located only 3½ miles from the village. The photograph below shows the cave in 1901, illuminated by magnesium flares. At that time, visitors carried candles and lamps for additional illumination. Today the cave is electrically lighted and the tour enhanced by the use of electronic narration. The trails have been equipped with buried loops of wire that transmit a radio message to the visitor who carries a small receiver. The narration provides the history, description and information about the cave as the visitor proceeds at his own pace. At each vantage point, the narration is individually transmitted, providing an unusual experience of modern technology in a setting that is eons old.

¶OPEN: all year 8:00 A.M. to 8:00 P.M. (or for groups by special arrangement) ¶SELF-GUIDED TOURS: Electronic hearing devices provided. Proceed at your own pace (approximately 30 minutes). Cave fully lighted ¶ON PREMISES: gift

Onyx Cave, Arkansas.

shop, Gay 90s Museum, picnicking ¶Nearby: all facilities ¶Nearby attractions: Eureka Springs Arts and Craft Village, Bible Museum, Passion Play of Christ, Saunder's Gun Museum, Pea Ridge Battlefield, Beaver Dam and Lake, Withrow Springs State Park, Cosmic Cavern, War Eagle Cave, Civil War Cave, Hurricane River Cave, Dogpatch Caverns, Bull Shoals Caverns, Buffalo National River.

WAR EAGLE CAVERN

Mailing address: P.O. Box 1381, Rogers, AR 72756   *Phone:* (501) 789-2909
Directions: ¼ mile off State Highway 12, halfway between Eureka Springs and Rogers.

War Eagle Cavern is located on a peninsula overlooking Beaver Lake. Locally known since Civil War times as Bat Cave, it was a cool location for parties and dances before the advent of air conditioned meeting halls. Today, the visitor parks on the hill above the cave and follows a trail to the entrance. The lake level is only a few feet from the entrance and it is possible to visit the cave by boat from the main lake. The broad, flat arch of the entrance as seen in the picture clearly shows the bedding plane of the limestone, and closer examination of the rock shows evidence of fossils characteristic of the region. The main passage follows a small stream that winds from the entrance back under the hill in a nearly horizontal direction. Several large domes occur in the ceiling. Rimstone pools are seen in the stream and, as the passage constricts, the tour ends and the visitor returns to the entrance. There is more cave passage than that

War Eagle Cavern, Arkansas.

exhibited, but the present trip covers only about ½ mile of trail. The management plans to extend the public route to include more of the cave at a later date.

¶OPEN: May 1 through October 31 daily 9:00 A.M. to 6:00 P.M., Sundays 1:00 p.m. to 6:00 P.M.   ¶GUIDED TOUR: 45 minutes   ¶ON PREMISES: gift shop and picnicking   ¶NEARBY: restaurant, motel, camping, trailer camp ¶NEARBY ATTRACTIONS: Eureka Springs, Onyx Cave, Cosmic Cavern, Civil War Cave, Beaver Lake, Diamond Cave.

CALIFORNIA

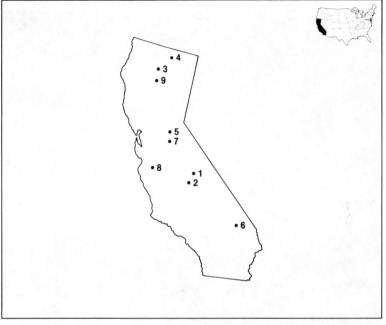

1 Boyden Cavern 2 Crystal Cave 3 Lake Shasta Caverns 4 Lava Beds National Monument 5 Mercer Caverns 6 Mitchell Caverns State Reserve 7 Moaning Cavern 8 Pinnacles National Monument 9 Subway Cave

CALIFORNIA HAS MORE THAN 800 miles of Pacific coastline and extends nearly 200 miles inland. Its land surfaces rise from 282 feet below sea level at Death Valley to 14,495 feet high atop Mount Whitney. Within this area are all the climatic conditions found in the rest of the United States, and all types of caves as well. Limestone caves, lava tubes, boulder caves, and sea caves are found in the various regions. Volcanic action 70 to 100 million years ago made spectacular scenery and mountain vistas, but conditions did not favor the formation of large cave systems and huge underground rooms. Though there are beautiful caves in the state, they are relatively small.

BOYDEN CAVERN

Mailing address: May to October — Box 817, Kings Canyon National Park, CA 93633; *Phone:* (209) 336-2391 November to May — Box 78, Vallecito, CA 95251 *Phone:* (209) 736-2798 *Directions:* Take California Route 180 north from the Kings Canyon National Park Visitors' Center at Gen. Grant's Grove about 18 miles to where highway crosses the Kings River.

Boyden Cavern, California.

Situated in the 8000-foot-deep Kings River Canyon, Boyden Cavern lies beneath the 2000-foot-high marble walls of Kings Gates. The cave is in marble bedrock (limestone which has been subjected to great heat and pressure) and contains interesting patterns and designs. It is well decorated, and there is sufficient surface water finding its way into the chambers to provide sparkling drops of water on small soda straw speleothems which give the cave scene brilliance and color.

¶OPEN: May to October 10:00 A.M. to 5:00 P.M. ¶GUIDED TOUR: 45 minutes ¶ON PREMISES: snack bar, gift shop, picnicking ¶NEARBY: camping and cabins ¶NEARBY ATTRACTIONS: Grizzly Falls, Roaring River Falls, Zumwalt Meadows, Sequoia National Park, Crystal Cave, hiking and horseback riding.

CRYSTAL CAVE

Mailing address: Sequoia National Park, Three Rivers, CA 93271 *Phone:* (209) 565-3341 *Directions:* State Route 198, approximately 55 miles northeast of Visalia.

It would be difficult to find a more impressive scenic drive than the 9-mile road that leads from the Giant Forest in Sequoia National Park to the parking lot near Crystal Cave. From here a self-guiding nature trail descends a half mile to the cave entrance in the bottom of the canyon of Cascade Creek. After crossing the creek on a rustic bridge, one sees an impressive series of waterfalls on either side. The opening of the cave is near the base of the lower falls. The entrance is equally impressive—a natural arch 30 feet wide and 16 feet high leading to a wrought iron gate that resembles a huge spiderweb. Exploration beyond this point is with a guide only.

The cave, formed in gray and white crystalline marble, was discovered by park employees after Sequoia Park had been created; protection of the delicate formations has been an important policy. This continuous care provides a rare view of completely undefiled cave architecture. The formations and dripstone deposits appear as they did when the cave was discovered. The trail leads through a number of narrow passages, descends into several large rooms, and then returns to the entrance. This cave experience makes a splendid counterpart to the Sequoia trees on the surface.

Crystal Cave, Sequoia National Park, California.

¶OPEN: daily mid-June to Labor Day; weekends May and September weather permitting, 9:30 A.M. to 3:00 P.M.   ¶GUIDED TOUR: 55 minutes (allow time to walk to the cave)   ¶ON PREMISES: no facilities at the cave except parking lot and rest rooms   ¶NEARBY: nearest facilities about 20 miles   ¶NEARBY ATTRACTIONS: Sequoia trees, Kings Canyon National Park, Boyden Cavern, Mount Whitney (highest point in California), Tule River Indian Reservation.

Lake Shasta Caverns, California.

LAKE SHASTA CAVERNS

Mailing address: P.O. Box 801, O'Brien, CA 96070   *Phone:* (916) 238-2341
Directions: About 13 miles north of Redding on Interstate 5. Turn off at Shasta
Caverns Road exit, about 2 miles to ticket office and catamaran dock.

Most cave trips are notable mainly because of the features within the cave.
Lake Shasta Caverns offers three experiences, each different and each worth
the trip. First the visitor is taken by catamaran ferry across an arm of Shasta
Lake. The stable and sturdy boat, aptly named *Cavern Queen*, makes the
15-minute crossing while passengers enjoy a fine view of the lake and pine-
covered shores. The second part of the tour affords visitors a breathtaking ride
by bus on a road which winds 800 feet above the lake to a small chalet built at
the entrance of the cave and clinging tenaciously to nearly vertical walls. The
third part is the cave trip itself. This finely decorated cave contains a spectacular wall of draperies and flowstone in the Cathedral room worthy of the
efforts expended to reach it. An additional highly decorated room will be
opened to the public soon. The return trip down the outside of the mountain
provides another view of the lake and wooded mountains. A booklet recount-
ing the early history of the cave is available for sale at the ticket office. This
efficiently operated tour is an outstanding attraction.

¶OPEN: all year 8:00 A.M. to 5:00 P.M., May through September. October
through April, three tours daily—10:00 A.M., 12:00 P.M., 2:00 P.M.   ¶GUIDED
TOUR: including bus and boat ride, 2 hours   ¶ON PREMISES: snack bar, gift
shop, picnicking   ¶NEARBY: all facilities, including fishing and boating at
Shasta Lake resort areas   ¶NEARBY ATTRACTIONS: Shasta Lake Recreation
Area, Shasta Dam, fishing, boating, swimming, Castle Crags State Park,
McArthur-Burney Falls Memorial and State Park.

LAVA BEDS NATIONAL MONUMENT

Mailing address: P.O. Box 867, Tulelake, CA 96134   *Phone:* (916) 667-2601
Directions: Off State Highway 139 about 20 miles south of Tulelake to
County 111.

More than 300 volcanic caves are known in this vast area of lava terrain, and
the lava tubes, blister cones, and other volcanic features are among the finest
examples in the world. Nineteen of the caves are open for exploration by visitors. At Lava Beds National Monument Headquarters a brochure is available
containing a map showing the locations of some of these caves. The brochure
also contains règulations for the protection of these caves and lists precautions
for visitors' safety. A loop drive near the Park Headquarters permits access to
19 caves, and only a short walk is necessary to reach the entrances. Catacomb
Cave, shown below, is named for the peculiar niches in the walls resembling the

Catacomb Cave, Lava Beds National Monument, California.

burial places of ancient Rome. This cave has 1½ miles of passage. Other caves in the loop drive are: Indian Wells Cave, Labyrinth Cave, Mushpot Cave, Thunderbolt Cave, Golden Dome Cave, Blue Grotto and a dozen more. Three other caves that can be visited are: Merrill Ice Cave, containing a frozen waterfall and river of ice that remains year round; Skull Ice Cave, the largest of the caves and in three levels; and Valentine Cave, which contains excellent examples of varying water flow levels. The caves are not lighted, but gasoline lanterns are available at headquarters. Visitors are asked to bring an extra source of light.

In addition to the remarkable geologic interest of the park, there is also historic interest, for the park was the site of the poignantly tragic Modoc War. In 1872, after several years of disputes with settlers, "Captain Jack" and his band of Modoc Indians took refuge in the lava beds immediately south of Tule Lake. In the area now known as Captain Jack's Stronghold, the small Modoc band held out against Federal and volunteer troops for nearly six months.

¶OPEN: all year 8:00 A.M. TO 5:00 P.M. ¶SELF-GUIDED TOURS: for as many hours as desired. Bring lights and borrow lanterns (free) at Headquarters. ¶ON PREMISES: camping, picnicking. From September 15 to May 15 water must be carried from Headquarters. ¶NEARBY: all facilities in Tulelake (30 miles away) ¶NEARBY ATTRACTIONS: Tulelake Wildlife Refuge, Medicine Lake Recreation Area, Klamath Lake, Clear Lake.

MERCER CAVERNS

Mailing address: P.O. Box 509, Murphys, CA 95247 *Phone:* (209) 728-2101 *Directions:* One mile north of Main Street, Murphys, State Highway 4, on Sheep Ranch Road.

The foothills of the Rockies near Sacramento are rich in the history of the Gold Rush of 1849, and there are many stories about the bizarre characters of those chaotic times. Every part of the country was scoured by prospectors seeking a bonanza. Even after placer gold was exhausted, die-hard enthusiasts searched the hills. One such prospector was Walter Mercer, who discovered a blowing hole and proceeded to excavate it. He found no gold, but instead found an unusual cave that he proceeded to develop and show as a tourist attraction. It was not an easy cave to develop. One side was nearly vertical and was a fault line in the limestone. This fault had caused a huge rift in the bedrock extending down nearly 250 feet into the earth. Mercer diligently worked to build trails and ladders in order to show this unusual pit. He exhibited it with torches and candles, and in 1887 (see photo) visitors donned overalls and gloves to make the five-hour trip. Today, the cave is electrically lighted; the stairs are sturdy though steep, and the formations are as pristine as when Mercer first entered. There are over 200 steps in the cave, but the view of a fine cluster of aragonite flowers at the lowest level is worth the effort.

Mercer Caverns, California.

¶OPEN: June to September 9:00 A.M. to 5:00 P.M. The rest of the year, weekends and school holidays 11:00 A.M. to 4:00 P.M.   ¶GUIDED TOUR: 35 minutes   ¶ON PREMISES: gift shop, picnicking   ¶NEARBY: restaurant, camping, motels, hotels, cabins, trailer camp   ¶NEARBY ATTRACTIONS: Calaveras Big Tree State Park, Columbia Historic State Park, Moaning Cavern, Angels Camp.

MITCHELL CAVERNS

Mailing address: P.O. Box 1, Essex, CA 92332   *Phone:* none   *Directions:* 16 miles north of Interstate 40 on Essex Road.

The Mojave Desert is a wild and barren place. Once the seasonal hunting grounds of the Chemehuevi Indians, later an area for silver prospectors, it is now an east-west corridor for traffic via modern highways. Prospective visitors can follow signs to Mitchell Caverns Natural Reserve in the Providence Mountains and will approach it with a view similar to that of the early prospectors. On the flank of the Providence Mountains, about 4000 feet above sea level, is a small stone house once occupied by Jack Mitchell and his wife, who developed the two little caves that now bear Mitchell's name. Acquired by the California State Parks Commission in 1956, these caves are now open to the public between October and June.

An easy trail skirts the slope of the mountains to the El Pakiva entrance.

Mitchell Caverns Natural Reserve, California.

Electric lights and concrete walks make it a comfortable guided tour. The cave is very dry, but not dusty, and the formations are clean and attractive. The addition of an artificial tunnel has connected the two caves so that exit is made through Tecopa Cave. A cliffside trail leads back to the visitors' center and another interesting view of the desert. Two nature trails are provided for hikers: the Mary Beal Nature Trail and the Crystal Spring Canyon Trail. With special prior arrangement, spelunking tours for experienced caving groups can be arranged to Winding Stair Cave, two miles away. This trip requires ropes and descends 350 feet in a series of free-fall drops that range from 50 to 140 feet.

¶OPEN: October to June 8:00 A.M. to 4:30 P.M. ¶GUIDED TOUR: 1 hour. Tours conducted at 1:30 P.M. daily; Saturday and Sunday at 10:00 A.M., 1:30 P.M., and 3:00 P.M. ¶ON PREMISES: camping, picnicking ¶NEARBY: restaurant, snack bar, gasoline (23 miles), motel, cabins (65 miles) ¶NEARBY ATTRACTIONS: (more than 30 miles away, but under 75 miles) Needles, Lake Mohave, Hole in the Wall Recreation Area.

MOANING CAVERN

Mailing address: Box 78, Vallecito, CA 95251 *Phone:* (209) 736-2708 *Directions:* 2 miles south of Vallecito on the Columbia-Sonora Road.

Located in the heart of the Motherlode Country, this cave was visited by prospectors in search of gold. The entrance and interior of the cave were also well known to early Indians; in fact the cave was used as a burial pit, for the Indians felt that spirits dwelt within (probably because of the moaning sound heard at the entrance that gives the cave its name). Present-day scientists believe that this is the sound of dripping water amplified by the hollow chambers in the rock. The cave is entered down a staircase within a small building serving as a gift shop and ticket office. A steeply descending crevice leads to a platform overlooking a remarkable room 125 feet high and more than a hundred feet across. The walls are decorated with pristine formations preserved through the years because of their inaccessability to explorers and visitors alike. In 1922 a circular staircase was constructed from the natural platform to the floor of the room. It is a marvel of engineering, but strenuous to climb. Visitors are encouraged to climb slowly, enjoy the scenery, and listen to the "moaning" of the cave. If it is any consolation to the weary visitor, the tour descends only half of the explored depth of the cave. The view from the top of the platform is worth the effort.

¶OPEN: all year 10:00 A.M. to 5:30 P.M. ¶GUIDED TOUR: 40 minutes ¶ON PREMISES: gift shop, camping, snack bar, picnicking ¶NEARBY: restaurant, motels, hotel, cabins, trailer camp ¶NEARBY ATTRACTIONS: Columbia Historic State Park, Calavaras Big Trees State Park, Mercer Caverns, Yosemite National Park (60 miles).

Moaning Cavern, California.

PINNACLES NATIONAL MONUMENT

Mailing address: Paicines, CA 95043   *Phone:* (408) 389-4578   *Directions:* Can be reached from south through King City, then take County Road G13 to Bitterwater, then north on State Highway 25.

The caves of beautiful Pinnacles National Monument are but a small part of the scenic value of the region. Nevertheless they are of interest, as they were formed by large boulders that have wedged in narrow canyons to bridge gorges. The caves are reached from several trails in Pinnacles National Monument of which Caves Trail and Moses Spring Trail give the best views. All trips are self-guided, and leaflets are available listing points of interest keyed to numbered stakes along the way.

¶Open: all year 7:00 a.m. to 8:00 p.m. daily   ¶Self-guided tours: Although the trails are lighted in most of the caves, flashlights are advisable.   ¶On premises: naturalist services, hiking, picnicking, camping, evening talks on summer weekends, small museum   ¶Nearby: nearest gasoline at Paicines (23 miles); no other facilities within 25 miles   ¶Nearby attractions: Bolado Park, Soledad Mission Ruins, Santa Lucia Memorial Park, Fremont Peak State Park.

SUBWAY CAVE

Mailing address: U.S. Forest Service, P.O. Box 218, Fall River Mills, CA 96028   *Phone:* (916) 336-5521   *Directions:* State Highway 89 and State Highway 44 at Old Station. Cave is located ½ mile north of information center on State Highway 89.

Subway Cave, a short hike from the Forest Service campground, is a sinuous lava tube with several entrances. A map posted at the entrance shows the twists and turns of this unlighted and natural cave. The floor of the cave is rocky but not slippery. There are some fine examples of lava features such as tiny "volcano-like" mounds in the floor, remnants of "shelving" of lava now solid stone, and "lavacicles," tiny points from the ceiling.

¶Open: all year, weather permitting   ¶Self-guided tours: Bring lights as the cave is not lighted.   ¶On premises: picnicking, camping   ¶Nearby: restaurant, snack bar, gift shop, trailer camp, motel   ¶Nearby attractions: Lassen Volcanic National Park, Lake Shasta Caverns, Eagle Lake, McArthur-Burney Falls State Park, camping at Hat Creek.

COLORADO

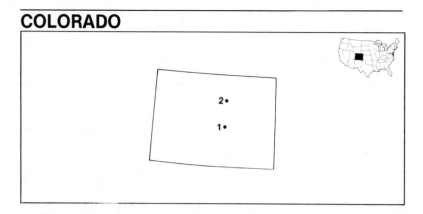

1 Cave of the Winds 2 Denver Museum of Natural History (artificial cave)

THE CENTRAL MOUNTAIN REGION OF COLORADO divides the state nearly in two and is a complex of many different kinds of rocks. At the time of the building of the Rocky Mountains 70 to 100 million years ago the rocks were strongly metamorphosed (altered by heat and pressure). The limestones in this complex were recrystalized and changed to marble by the heat and pressure and were sometimes elevated to great heights by mountain-building processes. The caves of the region, formed in remnants of soluble limestone, did not achieve the large size of caves in other parts of the country.

Most of the discoveries of the natural caves of this state came when prospectors where searching for gold and silver in the late 1800s.

CAVE OF THE WINDS

Mailing address: Box 228, Manitou Springs, CO 80829   *Phone:* (303) 685-5397
Directions: 6 miles west of Colorado Springs on U.S. 24, take Exit 50 off Interstate 25 west to Manitou Springs, ½ mile to cave.

Cave of the Winds, named for the moaning sound at the natural entrance, is perched like an eagle's nest high on the side of Williams Canyon. The Visitors' Center is also located precariously on the cliff face affording a breathtaking view. Within the cave all of the formations have ceased growing. The cave might have once been dusty, but now the floor is completely cemented. The trails and stairways are securely built; the lighting provides brilliantly colored effects. It is a crazily confusing and delightfully complex cave.

A word of caution to elderly or infirm persons. The cave is about 7000 feet above sea level. The stairs, which have 164 steps in all, require visitors to be in good physical shape, but ample time is permitted to climb them and to admire the scenery enroute. A drive between the cave and Manitou Springs via Williams Canyon can be an enjoyable sidelight of the trip.

¶OPEN: all year 8:00 A.M. to 10:30 P.M. in summer and from 9:00 A.M. to 5:00

P.M. in winter   ¶GUIDED TOUR: 40 minutes. In summer tours leave every 15 minutes. Special combination tickets are also available in summer, which include a spelunker's supper.   ¶ON PREMISES: restaurant, snack bar, gift shop ¶NEARBY: all facilities in resort area   ¶NEARBY ATTRACTIONS: center of resort and vacation area. Pikes Peak region, Colorado Springs, U.S. Air Force Academy, Garden of the Gods, Royal Gorge, Eleven Mile Reservoir State Recreation Area, Cripple Creek mining town, Fort Carson, Will Rogers Shrine of the Sun, Cheyenne Mountain, Zoo, Florissant Fossil Beds National Monument

DENVER MUSEUM OF NATURAL HISTORY (artificial cave)

Mailing address: City Park, Denver, CO 80205   *Phone:* (303) 575-3964
Directions: On State Highway 2 in Denver City Park next to the zoo.

In 1912 the Denver Museum fielded an expedition to Mexico, and scientists removed formations from two caves: El Potosi and Xochitl Caverns. The cave material, consisting of calcite, aragonite, and selenite crystals, was used to create a display in the museum to show how the cave rooms looked in their natural state. Half of the display, which exhibits calcite and aragonite crystals, is from El Potosi; the other half is from Xochitl Caverns and exhibits huge selenite crystals in two-foot by two-foot sections covering the display cave wall. Unfortunately, this type of exhibit seems to sanction cave vandalism and gives a false sense of intrinsic value to the formations. In their natural state, formations are alive and growing. Once removed they die and lose much of their beauty. It is very unfortunate that vandalism, under the guise of science, has destroyed two fine caves that required thousands of years to form.

¶OPEN: all year Monday through Saturday 9:00 A.M. to 4:30 P.M.; Sundays and holidays from noon to 4:30 P.M.

CONNECTICUT

THERE ARE NO LARGE CAVES IN Connecticut, only a few shelters and crevices. There are several limestone caves with formations, but none are open to the public.

DELAWARE

SITUATED ALMOST ENTIRELY IN THE Atlantic Coastal Plain, Delaware includes only a tiny area of the Piedmont Plateau where caves are found. There are no show caves open to the public.

DISTRICT OF COLUMBIA

1 Lincoln Memorial Underground (artificial cave)

THERE ARE NO NATURAL CAVES in Washington, D.C.

LINCOLN MEMORIAL UNDERGROUND (artificial cave)

Mailing address: National Capital Parks-Central, 900 Ohio Drive SW, Washington, D.C. 20242   *Phone:* (202) 426-6841   *Directions:* beneath the Lincoln Memorial building on the Mall, in downtown Washington.

The Lincoln Memorial, dedicated in 1922 by Warren Harding, has a labyrinth resembling a catacomb beneath the public area. This is not a natural cave, but soda straws formed by dripping water give that impression. These stalactites are formed by the minerals leached out of the cement and mortar of the structure and grow very differently from similar stalactites in caves. Cement is made from limestone that is crushed and burned. In the process the carbon dioxide in the stone is driven off. When water is added to make concrete, calcium hydroxide is formed, which absorbs carbon dioxide from the air and under the right conditions produces stalactites very rapidly. The Park Service conducts tours of the Lincoln Memorial Underground Monday through Friday at 7:30 P.M. and Saturday, Sunday, and Wednesday at 2:00 P.M. Tour are limited to 15 persons. Reservations should be made beforehand by letter or telephone; ask for the Mall Operation.

¶OPEN: all year   ¶GUIDED TOUR: 30 minutes. Bring flashlights and wear old clothes.

FLORIDA

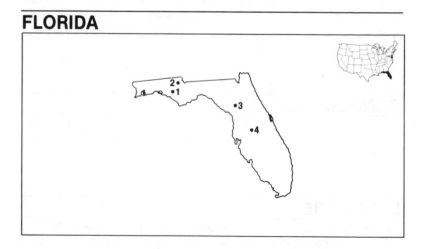

1 Falling Waters State Recreation Area 2 Florida Caverns State Park 3 Florida State Museum (artificial) 4 Walt Disney World (artificial)

THE ATLANTIC AND GULF COASTAL PLAINS, consisting of several thousand feet of sands, gravel, clay, and corraline limestone, underlie most of Florida. Although there are hundreds, perhaps thousands, of caves, most are now submerged and exist only as springs. At times during the ice ages, the caves were drained by the lowering of the sea level. When the caves were air-filled stalactites and stalagmites began to form; present-day divers have found secondary calcite formations far beneath the surface. Further evidence of climatic changes are the remains of mastodons and other extinct animals that have been found in the springs.

FALLING WATERS STATE RECREATION AREA

Mailing address: P.O. Box 660, Chipley, FL 32428   *Phone:* (904) 638-4030
Directions: Off Interstate 10, 3 miles south on State Highway 77A.

It is not possible to enter the cave at Falling Waters, but it is worth the short hike to see the 80-foot waterfall that drops into the sink entrance. The park, in hilly limestone country, has a number of sinkholes that are exhibited by means of an elevated walkway through an attractive wooded area. There are other caves in the 152-acre recreation site, but special permission is required in order to enter them.

¶OPEN: all year 8:00 A.M. to sundown   ¶SELF-GUIDED TOURS: about 1 hour
¶ON PREMISES: picnicking, swimming, camping   ¶NEARBY: restaurant, motel
¶NEARBY ATTRACTIONS: Florida Caverns, Gregory House, Three Rivers State Park, Gulf Coast resort area.

FLORIDA CAVERNS STATE PARK

Mailing address: 2701 Caverns Road, Marianna, FL 32446   *Phone:* (904) 482-3632   *Directions:* State Highway 167, 3 miles north of Marianna.

Florida Caverns, located near the highest point in Florida (345 feet above sea level), has sufficient elevation to allow the water that formed the cave to have drained away, providing a comfortable air-filled passage for visitors to explore. The caverns were well known to the Indians before the first visitation by the Spanish in 1693. Now a part of an attractive state park, the cave is just barely beneath the ground, and in some places there are only a few feet of limestone covering it. At the lowest places in the cave there are perennial pools that fluctuate with the water table. Geologically the cave is relatively young, possibly formed during the last Ice Age. The speleothems are remarkable in that they have formed from the relatively thin cover of limestone above. The 1700-acre park is of speleological interest, as the Chipola River flows underground for several hundred feet, providing the natural bridge that was used by General Jackson and his army in 1818. Several times a year, at high water, the level of the water will rise, partially flooding portions of the cave. At that time

Florida Caverns State Park, Florida.

the tour is shortened and another of the four entrances is used to gain access to the portion not submerged.

¶OPEN: all year 8:00 A.M. to sundown ¶GUIDED TOUR: 1 hour ¶ON PREMISES: camping, trailer camp, picnicking, swimming ¶NEARBY: restaurant, snack bar, motel, hotel, golf ¶NEARBY ATTRACTIONS: Jim Woodruff Dam, Falling Waters State Park, Battle of Marianna State Historical Memorial, Three Rivers State Park, Lake Seminole, Seminole State Park (Georgia).

FLORIDA STATE MUSEUM (artificial cave)

Mailing address: Museum Road, University of Florida, Gainesville, FL 32611
Phone: (904) 392-1721 *Directions:* Take State Highway 26 east off Interstate 75 at Gainesville.

This walk-through exhibit is called simply "A North Florida Cave." Located on the University of Florida campus, the museum features natural and social history.

¶OPEN: all year Monday through Saturday 9:00 A.M. to 5:00 P.M.; Sunday 1:00 P.M. to 5:00 P.M. ¶SELF-GUIDED TOUR: about 15 minutes ¶ON PREMISES: gift shop ¶NEARBY: all facilities in Gainesville ¶NEARBY ATTRACTIONS: Manatee Springs State Park, Floating Islands, Magnolia Lake State Park, Gold Head Branch State Park.

WALT DISNEY WORLD (artificial caves)

Mailing address: P.O. Box 40, Lake Buena Vista, FL 32830   *Phone:* (305) 824-2222   *Directions:* Interstate 4 about 20 miles south of Orlando.

There is no need to elaborate on the remarkable development of this private park, the dream of Walt Disney, and the creation of this mind-boggling illusion built in the scrub-covered savannah of central Florida. Disney's mind roamed over all of the facets of life and he did not neglect caves, which (although an infinitesimal part of this park) he scrupulously duplicated and presented in several displays.

Huck Finn's Cave, on Tom Sawyer Island, is an artificial tunnel and maze that gives the feeling of the actual Mark Twain Cave in Missouri. Fabricated of sprayed concrete and wire mesh, Huck Finn's Cave twists and turns in a confusing fashion, traversing an island and providing fun for the child in all of us. While geologically impossible in Florida, the illusion will satisfy even the most discerning. Pirate Cave, part of the Pirates of the Caribbean, Adventure Land display, is only glimpsed briefly as the conveyance that carries the visitor glides through the entrance of the pirates' lair. It is a stage set, designed to be seen only from one camera angle; but again, the illusion of an actual cave is excellent. Probably no one visits Walt Disney World just to see the caves, but while there, it is worthwhile seeing these careful reproductions.

¶OPEN: all year. Magic Kingdom open daily 9:00 A.M. to 7:00 P.M. and to 1:00 A.M. in summer   ¶TOURS: Huck Finn's Cave, self-guided; Pirate Cave is part of a motorized ride.   ¶ON PREMISES: all facilities for a complete vacation holiday   ¶NEARBY ATTRACTIONS: Walt Disney World has become a point of destination in Central Florida.

GEORGIA

GEORGIA IS DIVIDED DIAGONALLY into two regions: to the southeast is the Atlantic Coastal Plain, and to the northwest is the Appalachian Piedmont Plateau. Only in the extreme northwest tip of the state are there mountains. There are many caves in the Piedmont highland area. Although several have been opened in the past as show caves, at this time there are none open.

HAWAII

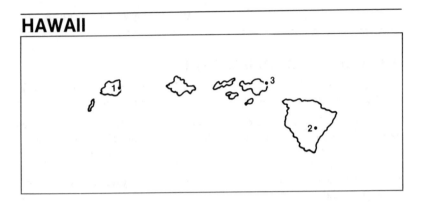

1 Fern Grotto (Kauai) 2 Thurston Cave (Hawaii) 3 Waianapanapa Cave (Maui)

ALL OF THE NINE MAJOR ISLANDS of the state of Hawaii are of volcanic origin and are peaks of a submerged ridge. The northwest-trending ridge divides two oceanic deeps, both of which descend more than 18,000 feet below sea level. The highest lone peaks are more than 13,000 feet above sea level, so that the total relief is 31,000 feet — more than six miles. The big island of Hawaii has two volcanos, Mauna Loa and Kilauea, that have erupted in the past ten years. When active the lava from the central caldera flows through lava tubes that extend off the flank of the dome-shaped volcanoes. Sometimes these tubes reach the sea and it is possible to see the flowing lava racing along underground through ceiling openings in the cooling surface lava. Numerous lava tubes exist on the main islands. The older islands to the west have coral reefs and coastal volcanic rock that has been eroded by the waves. Sea caves, arches, and blowholes are common. Well known by the sea-going Polynesians, these caves were named, visited, and sometimes revered as sanctuaries. A number of them were used as burial places.

FERN GROTTO

Mailing address: Wailua River State Park, Kapaa, Kauai, HA 96746   *Directions:* On the island of Kauai, about 2 miles south of Kapaa on State Highway 56.

The Wailua River Reserve is a 297-acre area extending along the north and south forks of the Wailua and has been set aside as a park area to preserve the natural beauty of the river. The Fern Grotto area is about three miles upstream from the mouth of the river on the south fork. River boats from the Wailau Marina take visitors to Fern Grotto landing. From here it is a short walk through dense jungle to the grotto—a cave with an entrance lush with ferns hanging from the top of the opening. The grotto serves as an amphitheatre where Hawaiian singers and dancers give performances. There is little more to the cave than the shallow entrance, but the tropical and exotic setting makes it a point of destination for many visitors.

¶OPEN: daily all year during daylight hours   ¶SELF-GUIDED TOUR: The Fern Grotto boat trip is a private concession and a tour of the river includes the performances at the grotto. The cave does not require lights.   ¶ON PREMISES: There are no facilities at the grotto.   ¶NEARBY: camping, restaurant, boating and hiking   ¶NEARBY ATTRACTIONS: Kokee State Park, Na Pali Coast State Park, Barking Sands Beach, Opaaikaa Falls, Waimea Canyon State Park.

THURSTON CAVE

Mailing address: Hawaii Volcanoes National Park, HA 96718   *Directions:* On State Highway 11 about 20 miles south of Hilo.

The National Park Service maintains Thurston Lava Tube, one of hundreds of lava tubes on park lands, as a visitor attraction. Formed in one of the Kilauea Volcano eruptions, the cave has cooled and now provides an easy walking tour of a gently sloping tunnel. Vegetation has returned to the region after the fire scorching of the volcano, giving a picturesque setting of vines and ferns around the cave entrance. An interpretive program is available at the cave which identifies outstanding natural features.

¶OPEN: all year during the daylight hours   ¶SELF-GUIDED TOUR: Cave is located a short walk from the parking lot. Does not require lights although a flashlight is recommended.   ¶ON PREMISES: picnicking   ¶NEARBY: camping, restaurant, hotel   ¶NEARBY ATTRACTIONS: Kilauea Volcano, Mauna Loa volcano, City of Refuge National Historical Park, Akraka Falls State Park, Hapuna Beach, Lava Tree State Monument, Wailoa River State Recreation Area.

WAIANAPANAPA CAVE

Mailing address: Waianapanapa State Park, Hana, Maui, HA 96713   *Directions:* On the island of Maui off State Highway 36 via Honokalaui Road, ½ mile east of Hana Airport junction.

A side road off Highway 36 leads to Waianapanapa State Park where two caves, formed by the partial collapse of a lava tube, are located. The caves, composed of lava from Haleakala Volcano, are close together by the sea and called Waianapanapa Cave and Waiomao Cave. At Waianapanapa Cave it is possible to swim in a freshwater pool at the entrance and even beyond with flashlights or other lighting. There are legends concerning the caves. One tells of a princess who was slain deep inside Waianapanapa Cave. One can swim to a rock ledge where she was believed to have hidden. Every April, the month she was slain, the cave water, according to legend, turns red from her blood. Waiomao Cave does not receive direct flow, and the water is not fit for swimming.

¶OPEN: all year during daylight hours   ¶SELF-GUIDED TOUR: Access to the cave might require swimming. Bring lights and proper equipment.   ¶ON PREMISES: Cabins, camping, picnicking, swimming, fishing, hiking ¶NEARBY: restaurants, motel. Reservations necessary for cabins and camping. Write to: Division of State Parks, P.O. 1049, Waliluku, Mauai, HA 96793   ¶NEARBY AT-TRACTIONS: Iao Valley State Park, Poli Poli Springs, Halekii-Pihana State Monument.

IDAHO

1 Craters of the Moon National Monument 2 Crystal Ice Caves 3 Idaho's Mam-moth Cave 4 Minnetonka Cave 5 Shoshone Indian Ice Caves

THE ROCKY MOUNTAINS dominate the western and northern skyline of this state. Volcanic activity 60 million years ago raised these massive mountains, leaving a rugged and irregular landscape. Early settlers who crossed the great western plains in the middle of the last century struggled over the volcanic wastelands formed by lava flows at the base of the mountains. Hundreds, perhaps thou-sands, of lava tubes can be explored in the flows, and several are open to the public.

CRATERS OF THE MOON NATIONAL MONUMENT

Mailing address: Arco, ID 83213   *Phone:* (208) 527-3257   *Directions:* U.S. Route 20/26, and 93A.

Part of this 83-square-mile park is designated as the Cave Area. In the central portion of this astonishing lava landscape, the Cave Area is reached by a half-mile walk from Loop Drive, a seven-mile road traversing the major portion of the tourist area. The five major caves in this area are Indian Tunnel, Boy Scout, Beauty, Surprise, and Dew Drop caves. Of these, Indian Tunnel is the largest, over 830 feet long. No lanterns are necessary in this lava tube because the tunnel ceiling has collapsed in several places. Boy Scout Cave has a floor of ice, even in summer. Lights or lanterns are necessary in the smaller caves, which are not lighted. Signs explain the interesting features along the trail. Close to the southernmost area is Great Owl Cave, away from the Cave Area. The entrance is ¾ mile from the parking lot. It is a lava tube about 500 feet long, 40 feet high, and 50 feet wide. A stairway leads into the cavern, and lights are necessary.

¶OPEN: all year, weather permitting   ¶SELF-GUIDED TOUR: from 1 to 2 hours. Bring flashlights or lanterns.   ¶ON PREMISES: snack bar, camping, picnicking, many exhibits   ¶NEARBY: restaurant, motels (18 miles east, 24 miles west) ¶NEARBY ATTRACTIONS: U.S. Atomic Energy Commission Reservation, Blizzard Ski area, Lava Lake, Shoshone Indian Ice Caves, Mammoth Cave, Fish Creek Reservoir.

CRYSTAL ICE CAVE

Mailing address: 514 Highland Street, American Falls, ID 83211   *Phone:* (208) 226-2465   *Directions:* 29 miles northwest of American Falls. Exit Interstate 15W at American Falls, cross the New Snake River Bridge on State Highway 39, north to Pleasant Valley Road.

Crystal Ice Cave is not like the more common lava tubes that occur in the lava flows in nearby Craters of the Moon National Monument. It is part of the Great Rift, a natural fissure in the rock, extending from Craters of the Moon to the Wapi Lava Flow, a distance of more than 43 miles. This rift, which has been explored to a depth of more than 800 feet, is the largest volcanic rift on the North American continent. It contains volcanic phenomena such as explosion pits, spatter cones, and lava dikes in a concentration seen in few other places on earth.

Developed by James Papadakas in 1961, the tour traverses a 1200-foot tunnel dug in the volcanic rock parallel to the rift, descending without steps to a level 160-feet beneath the surface. The ice formations are viewed through insulated windows to protect the ice. At the Hanging Balcony an unobstructed view of

Crystal Ice Cave, Idaho.

the ice formations is possible for 150 feet in both directions. The cave temperature is a constant 32° F, so it is advisable to bring a jacket.

¶OPEN: May 1 to October 31, 8:00 A.M. to 8:00 P.M.   ¶GUIDED TOUR: 1 hour ¶ON PREMISES: gift shop, trailer park, snack bar, picnicking   ¶NEARBY: all facilities in American Falls   ¶NEARBY ATTRACTIONS: Craters of the Moon National Monument, American Falls Dam and Reservoir, Maccare Rock, Register Rock State Park, Indian Springs, Walcott State Park, Lake Walcott, Aberdeen Sportsmen's State Park, Snake River, Idaho State University, warm springs, bathing, fishing, boating, old Fort Hall.

IDAHO'S MAMMOTH CAVE

Mailing address: P.O. Box 648, Shoshone, ID 83352   *Phone:* (208) 886-2684 *Directions:* State Highway 75, about 7 miles north of Shoshone, 1½ miles on private road.

Idaho's Mammoth Cave, Idaho. Photo by Mull.

This impressive lava tube developed by Richard and William Olsen was first visited in the early 1800s. Today visitors purchase a ticket, are handed a gasoline lantern, and are told: "The lights are on in the cave—just follow the trail." Lava tubes, formed by rivers of molten lava that have "frozen" on the outside and drained out of the inside, have a singularly dull floor plan. They normally consist of a subway-like tunnel with only slight changes in direction. There is a very limited chance of becoming lost. Cling securely to the gasoline lantern, however, for it lights the volcanic black passageways, casting interesting shadows. The passageway is large by lava tube standards, and lime minerals have leached through the lava, providing portions of the cave with light walls. This is unusual, as most lava tubes are almost totally black.

While it would be difficult to damage this type of cave, it can be defaced by littering or writing on the walls. A self-guided tour is a privilege; future visitors will appreciate your leaving the area as you found it.

¶OPEN: April through November 1, 8:00 A.M. until sundown ¶SELF-GUIDED TOUR: takes about 30 minutes. Cave is lighted with electric lights. ¶ON PREMISES: snack bar, gift shop, camping, trailer camp, picnicking, museum ¶NEARBY: restaurant, motel, cabins ¶NEARBY ATTRACTIONS: Shoshone Indian Ice Caves, Sun Valley, Sawtooth Recreation area, Craters of the Moon National Monument, Crystal Ice Caves.

MINNETONKA CAVE

Mailing address: USDA—Forest Service, 431 Clay Street, Montpelier, ID 83254 *Phone:* (208) 847-0375 *Directions:* 11 miles west of Montpelier, about 9 miles up St. Charles canyon.

The only limestone cave open to the public in Idaho is administered by the U.S. Forest Service under the Visitor Information Service program. Discovered by a hunter in 1906, it was known as Porcupine Cave for many years until it was developed as a project of the Works Progress Administration in 1938. Since that time electric lights have been installed, and modifications to the trail and surface area make the cave location available to thousands. The entrance elevation is approximately 7700 feet above sea level, making this one of the highest show caves in the country. A scenic foot trail leads from the parking lot to the cave entrance. The tour covers about a half mile of passageway consisting of a series of chambers, enlarged primarily by the breakdown of the distinctly bedded limestone. One of the rooms, 300 feet in diameter and 90 feet high, is most impressive. The formations are mostly dry, but some are active and are the source of the cave's Indian name, which means "Falling Waters." Because of the elevation, a leisurely pace is recommended both in the cave and on the trail. A jacket is suggested as the cave is a constant 40° F.

¶OPEN: June 15 to Labor Day, 10:00 A.M. to 5:00 P.M. ¶GUIDED TOUR: 1 hour

¶ON PREMISES: no facilities   ¶NEARBY: camping and picnicking. Nearest facilities in Montpelier (20 miles)   ¶NEARBY ATTRACTIONS: Bear Lake State Park, Bear Lake, Home Canyon Ski Area.

SHOSHONE INDIAN ICE CAVES

Mailing address: Box 122, Shoshone, ID 83352   *Phone:* (208) 886-7728
Directions: 17 miles north of Shoshone on State Highway 75.

Shoshone Indian Ice Caves are located in an extensive lava tube which serves as a natural icebox and at one time served also as a skating rink (see photo). While the main cave is now only a thousand feet long, it is the remnant of what was originally a 5-mile-long lava tube. It is unique in that the ice that forms in it during the winter survives during the summer. When the cold winter air returns, it freezes the water accumulated over the summer, thus producing a perpetual glacier effect. A visit to this cave will show the classic ice cave cycle at work. A fine trail circles the lava tube before visitors descend. It is cool underground; jackets are recommended.

Shoshone Indian Ice Caves, Idaho.

¶Open: May 1 to October 1, 8:00 a.m. to 7:15 p.m.   ¶Guided tour: 1 hour ¶On premises: gift shop, ten-ton statue of Shoshone Indian Chief Washakie ¶Nearby: restaurant, snack bar, camping, picnicking. Nearest motels, 15 miles   ¶Nearby attractions: Idaho's Mammoth Cave, Sun Valley, Magic Reservoir, Shoshone Falls, Craters of the Moon National Monument (40 miles).

ILLINOIS

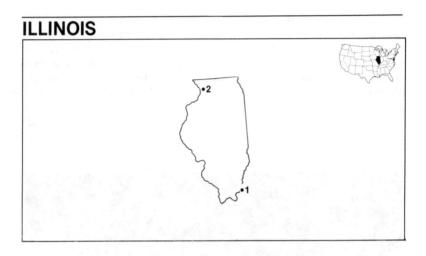

1 Cave-In-Rock State Park 2 Mississippi Palisades State Park

The northern Mississippi Valley plains were the terminus for the great glacial ice packs in recent geologic time, perhaps as little as 10,000 years ago. The accumulation of ice, possibly a mile thick, moved across the land surface and transported debris over great distances. When the ice retreated it deposited this material in moraines in random locations, thereby changing the local topography. Some of the caves formed by this melt water were filled with the debris carried in the ice. Only a few caves are found in this region, none of them show caves with guide service. The two caves that can be visited are on state park lands and each is on the banks of a great river although at opposite ends of the state.

CAVE-IN-ROCK STATE PARK

Mailing address: Box 338, Cave-In-Rock, IL 62919 *Phone:* (618) 289-4325 *Directions:* One-half mile east of State Highway #1 near Cave-In-Rock.

Cave-In-Rock State Park is a narrow strip of land bordering on the Ohio River for nearly a mile. The feature of the park is the large arched opening shown below and the cave that extends back into the limestone bluff. Perhaps no other cave in the United States has such a notorious history. First described by the French in 1729, it became a lair for outlaws in 1797. Situated in a strategic position to lure and trap river travelers, this cave became a den for several generations of scamps and worse. Among the early robbers was Samuel Mason, an ex-officer in George Washington's army. He was followed by the infamous Harpe brothers, whose perverted murders struck terror in the people of this entire country. The cave was later used as a hideout by a gang of counterfeiters who operated there until 1831.

Today the park is a popular spot for picnic parties and can be visited without lights during daylight hours. The rear of the 160-foot-deep passage has an upper opening that permits light to illuminate the rear of the cave. The only change that has occurred is that today, at flood stage of the Ohio River, it is possible to enter by boat. At pool stage, the river is deep enough for the popular passenger boat, the Delta Queen, to debark visitors on the shore for a tour of the cave.

¶OPEN: all year 8:00 A.M. to 10:00 P.M., weather permitting ¶SELF-GUIDED TOUR: The large cave entrance lights most of the cave in daylight. Bring lights for night exploration. ¶ON PREMISES: restaurant, snack bar, gift shop, camping, trailer camp, picnicking, swimming and boating ¶NEARBY: motel, hotel and cabins ¶NEARBY ATTRACTIONS: "Garden of the Gods," Shawnee-town, Lake Glendale Beach and Picnic Ground, Dam Village (Kentucky), Ohio River.

Cave-In-Rock, Illinois.

MISSISSIPPI PALISADES STATE PARK

Mailing address: P.O. Box 364, Savanna, IL 61074   *Phone:* (815) 273-2731
Directions: State Highway 84 on Mississippi River, 3 miles north of Savanna.

This park has two caves, Bob Upton's Cave and Bat Cave. Neither one is of great size but both are of historical interest in the region. Bob Upton, for whom the larger cave was named, was a resident of the area. In 1832, at the outbreak of the Black Hawk War, the settlers sent the women and children to Galena for safety, while the men remained to work the crops. One day a band of Indians appeared, and all the men except Upton fled to the river, where they had boats. Upton hid in the cave, and when the boats came near him he joined the other men. The cave has borne his name ever since. The entrance is in a bluff overlooking the Mississippi River.

¶OPEN: all year, daylight hours   ¶SELF-GUIDED TOUR: Follow trail signs in the park. No lights necessary   ¶ON PREMISES: camping, trailer camp, picnicking ¶NEARBY: restaurant, motels   ¶NEARBY ATTRACTIONS: Charles Mound (highest point in Illinois), Apple River Canyon, Grant Home, Plum Tree Winter Sports Area, Lake Le-Aqua-Na State Park.

Bob Upton's Cave, Illinois, Mississippi Palisades State Park.

INDIANA

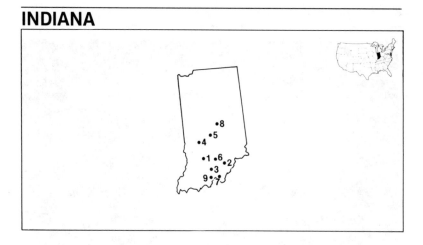

1 Bluespring Caverns 2 Cave River Valley Park 3 Marengo Cave 4 McCormick's Creek State Park 5 Porter's Cave 6 Spring Mill State Park 7 Squire Boone Caverns 8 The Children's Museum (artificial cave) 9 Wyandotte Cave

THE MISSISSIPPI VALLEY PLAINS are surfaced with rock waste and eroded soils from both the Rocky and Appalachian mountains. The material, brought by streams, the wind, and glaciers, deposited a layer of sediment which has produced one of the most fertile regions on earth. At times in the geologic past this area was submerged and formed an inland sea, producing the great limestone deposits of Indiana, Kentucky and Missouri. Southern Indiana contains some of the longest caves in the country. The nearly level bedding has produced large meandering passages with little change in elevation. The wooded hill country has many "closed" valleys that shelter the entrances of caves. Waterfalls and disappearing streams are common in the forested limestone section of the state.

BLUESPRING CAVERNS

Mailing address: R.R. 11, Box 479, Bedford, IN 47421 *Phone:* (812) 279-9892
Directions: About 6 miles south of Bedford off U.S. 50, 1½ miles west of State Highway 37.

The limestone in central Indiana has been little disturbed by earth-building processes. The bedding still shows the level sedimentary layers originally laid down. In Bluespring Caverns only a small part of the drainage system, which has been explored for more than twenty miles, is shown to visitors. Portions of this great cave have been known since the 1800s, but it was not until 1940 that the area now shown was discovered in a dramatic way. A large pond on the George Colgazier farm disappeared one night in a heavy rain, to reveal the entrance that is used today. Exploration of this new area was made along a water passage, but in 1971 trails, boats, and lights were installed to permit visitation by the public. A

Bluespring Caverns, Indiana.

short walk from the Visitors' Center takes the visitor into the sinkhole entrance room where he boards flat-bottomed boats for a 4000-foot ride — one of the longest underground boat rides in the United States. In the clear water one can see blind white fish and crayfish, the strangely adapted inhabitants of this subterranean world. The boat glides silently over the water, the only sounds being the soft popping and booming made when the water becomes trapped in small overhangs along the route.

¶OPEN: May 1 to September 30, daily, 9:00 A.M. to 6:00 P.M. Weekends, March and April, October and November   ¶GUIDED TOUR: 1 hour   ¶ON PREMISES: gift shop, camping and picnicking   ¶NEARBY: restaurant, motel   ¶NEARBY AT-TRACTIONS: Ed's Ghost Town, Lost River, Spring Mill State Park, Lake Monroe, Porter's Cave, Osborn Spring Park.

CAVE RIVER VALLEY PARK

Mailing address: Campbellsburg, IN 47108   *Phone:* (812) 755-4227   *Directions:* 2 miles north of Campbellsburg off State Highway 60.

Cave River Valley is a private park sheltered in a gorge in the hills of southern Indiana. Selected by an early settler as the "perfect mill stream," it exists today as an unspoiled bit of wilderness with dozens of caves, some of them unexplored. There are eight caves open to the public, and each requires a different degree of skill to explore. Before entering, it is necessary to register at the office where a description of the skills and equipment required is provided. River Cave, only a few yards from the picnic area, has an active stream that provided the power for the first grist mill built on the spot in 1817. Lights are required to enter, but it is an easy and interesting trip, and visitors feel a real spirit of exploration while splashing through the water of the underground stream. Endless Cave, just a hundred yards northwest of River Cave, is easily traversed and again requires flashlights after the first few feet. The other caves — Bear Den Cave, Crystal Spring Cave, Lake Cave, Frozen Waterfall Cave and Lamplighter's Cave — provide a treat for the more advanced spelunker . . . and these are only a few of the many caves in the park! For most people, the two main caves will provide sufficient adventure. Other attractions include fine trout fishing and hiking on trails through the beautiful scenery of this section of Indiana. As with all self-guided activities, it is hoped that those who visit this unspoiled area both above and below ground will respect the natural conditions.

¶OPEN: all year   ¶SELF-GUIDED TOUR: All of the caves require lights; guides can be provided by prior arrangement.   ¶ON PREMISES: snack bar, gift shop, camping, trailer camp, picnicking, trout fishing, 120-year-old cabin   ¶NEARBY: motel, restaurant within 3 miles   ¶NEARBY ATTRACTIONS: Spring Mill State Park, Marengo Cave, Wyandotte Caves, Bluesprings Cave, Devils Backbone, French Lick.

MARENGO CAVE

Mailing address: P.O. Box 217, Marengo, IN 47140   *Phone:* (812) 365-2705   *Directions:* Junction of Indiana State Rd. 64 and Indiana State Rd. 66, 9 miles north of I-64 via State Road 66.

Located in the center of the finest cave region of Indiana, Marengo Cave has

Marengo Cave, Indiana.

attracted visitors for nearly a hundred years. Discovered in 1883 by two children, brother and sister, the cave was developed the same year. The present entrance leads down a stairway into the main corridor, a huge quarter-mile tunnel, gracefully arched and showing the action of the water that formed it. The main passage is so massive that by special arrangement, troops of boy or girl scouts may stay there for an overnight camping adventure. The experience includes an evening tour of the undeveloped parts of the cave and a program giving information about the cave development and cave life. The regular tour includes a visit to a most highly decorated series of rooms where it is necessary to squeeze between huge stalagmites that have formed so close together that they nearly fill the passage. Most of the cave is dry, but the Crystal Palace area has many active formations.

¶OPEN: all year. Winter 9:00 A.M. to 5:00 P.M. Summer 9:00 A.M. to 7:00 P.M. ¶GUIDED TOUR: 1 hour   ¶ON PREMISES: restaurant, snackbar, gift shop, camping, trailer camp, picnicking, swimming pool ¶NEARBY: motel ¶NEARBY ATTRACTIONS: Wyandotte Cave, Squire Boone Caverns, Cave River Valley park, Crawford State Forest, Spring Mill State Park, Horseshoe Bend Marina, Corydon (first state capitol).

McCORMICK'S CREEK STATE PARK

Mailing address: Spencer, IN 47460 *Phone:* (812) 829-2235 *Directions:* State Highway 46, about 2 miles east of Spencer.

Among the many natural features of this park are two small caves open to visitors. A trail map and leaflet obtainable at the park entrance give information regarding their location. They are Sunken Cave and Wolf Cave, which are located at the northern end of this 1225-acre park.

¶OPEN: all year 7:00 A.M. to 11:00 P.M. ¶SELF-GUIDED TOUR: flashlights required ¶ON PREMISES: restaurant, snack bar, gift shop, camping, hotel, primitive cabins, trailer camp, picnicking, naturalist service, swimming pool, horseback riding, natural trails, playground, sports fields, trailside museum, campfire programs, stone quarry ¶NEARBY: motel, other facilities in Spencer ¶NEARBY ATTRACTIONS: Brown County State Park, Cagles Mill State Park and Dam, Cataract Lake, Monroe Reservoir, Lieber State Park, Franklyn College, Porter's Cave, goldfish hatchery.

PORTER'S CAVE

Mailing address: Route 1, Box 125, Paragon, IN 46166 *Directions:* About 45 miles southwest of Indianapolis off State Highway 67, on Interstate 70 at Little Point Exit.

Many waterfalls are found in the hills of south central Indiana and these attracted millers in the early years of the nineteenth century. One such miller was Francis K. Porter, who homesteaded land containing a picturesque waterfall that emerged from a cave. Today the grist mill is gone, the evidences of the homestead are overgrown, and the land has reverted to the condition in which Porter must have found it when Indiana became a state. The cave also is unchanged, and while a trail has been put within it, there are no electric lights.

Porter's Cave, Indiana.

Equipped with lantern or flashlight, the visitor follows an easily walked passageway alongside a stream deep within the mountain.

¶OPEN: March to November, dawn to dusk   ¶SELF-GUIDED TOUR: Bring your own lights and allow 35 to 40 minutes.   ¶ON PREMISES: restaurant, dining hall,

concession stand, gift shop, camping, cabins, trailer camp, picnicking, audi-
torium (90' × 60'), 100-bed lodge with kitchen facilities, softball field ¶NEARBY
ATTRACTIONS: Lieber State Park, Cataract Lake, McCormick Creek State Park,
Brown County State Park, goldfish hatchery.

SPRING MILL STATE PARK

Mailing address: Mitchell, IN 47446   *Phone:* (812) 849-4129 *Directions:* Off
State Highway 60, 3 miles east of Mitchell.

There are a number of caves in this very attractive 1319-acre state park. They are
available only after registering at the office where a guide or naturalist can ac-
company visitors to one or several. One of these, Upper Twin Cave, includes a
boat ride. Donaldson's Cave and Bronson's Cave can be visited free with a
naturalist who gives an interpretive and educational program for visitors.
Hamer Cave is not open to visitors, as it powers the park grist mill. For all of the
caves, bring lights and wear rugged shoes. Special permission is needed to visit,
and prior contact with the naturalist's office is recommended (812) 849-2206.
Some of the caves connect, and Upper Twin Cave is the source of the water
which travels through a karst window area to Lower Twin. There is another
karst window of roof collapse here, as the water flows into Bronson's Cave and
then out Donaldson's Cave. Even though they have different names, these are
part of one system which has been declared a National Natural Landmark. It is
in this system also that Dr. Carl Eigenmann did much of his outstanding
research on the blind fish which can be found in the cave water.

Donaldson's Cave, Spring Mill State Park, Indiana.

¶OPEN: Park open all year. Boat ride April to October 9:00 A.M. to 5:00 P.M. ¶GUIDED TOUR: 20 minutes. Boat traverses Upper Twin Cave for about 500 feet.   ¶ON PREMISES: all facilites including swimming pool, horseback riding, Pioneer Village, Grissom Memorial. Also Spring Mill Inn (reservations, call [812] 849-4081) and campground (reservations, call [812] 849-4138 April 1 to October 15)   ¶NEARBY ATTRACTIONS: Cave River Valley Caves, Marengo Cave, Wyandotte Cave, Devils Back Bone, Brown County State Park, Lake Greenwood, French Lick Mineral Springs.

SQUIRE BOONE CAVERNS

Mailing address: P.O. Box 411, Corydon, IN 47112   *Phone:* (812) 732-4381
Directions: 10 miles south of Corydon, on State Highway 135, then 3 miles east on Squire Boone Road.

One of the most famous men to be attracted to southern Indiana was Squire Boone, brother of Daniel Boone. Though not as well known as his brother, he was nevertheless a remarkable man of many talents and accomplishments. After an illustrious career as minister, hunter, Indian fighter, politician, and explorer, he settled down near a cave spring and built the first mill in Harrison County. The water that came from the cave spring entrance powered the mill race, but whether Squire Boone ever explored the cave beyond the entrance-way is not known. In 1973, two hundred years after the discovery of the cave, it was opened commercially by the construction of a tunnel into the side of the hill above the old mill site. This provided easy access to the cave stream below. The tour today provides a dazzling display of natural waterfalls that have been spanned with steel bridges and elevated platforms permitting the visitor to traverse the cave without discomfort. The rushing water adds movement to what could otherwise be a static scene. The trail follows the stream, which at one point plunges down a hole in the floor. The visitor is treated to an exciting view of these falls as he stands on a grating directly over them. It is not necessary to retrace any steps, as the tour ends at a spiral staircase leading back to the Visitors' Center.

¶OPEN: June to August, daily 9:00 A.M. to 6:00 P.M.; March to May and September to November 10:00 A.M. to 5:00 P.M.; December to February, weekends or by appointment   ¶GUIDED TOUR: 1 hour   ¶ON PREMISES: restaurant, snack bar, gift shop, camping, pioneer village, log cabins and craft shops, picnicking   ¶NEARBY: motel, trailer camp, restaurant   ¶NEARBY ATTRACTIONS: Marengo Cave, Wyandotte Cave, first state capitol (Corydon), Fort Knox (Kentucky).

Squire Boone Caverns, Indiana.

THE CHILDREN'S MUSEUM (artificial cave)

Mailing address: 3000 N. Meridian Street, Indianapolis, IN 46208   *Phone:* (317) 924-5431   *Directions:* Take 30th Street Exit off U.S. Highway 31 in downtown Indianapolis.

Called simply "An Indiana Cave," this artificial exhibit, completed in 1976, represents a limestone cave. Children and adults enter a maze-like corridor for more than 120 feet, where synthetic stalactites, stalagmites, and flowstone provide an underground experience exhibiting cave life and cave scenery in a simulated environment.

The Children's Museum, Indiana.

¶OPEN: all year, Tuesday through Saturday, 10:00 A.M. to 5:00 P.M.; Sunday, 1:00 P.M. to 5:00 P.M.   ¶SELF-GUIDED TOUR: Passage is lighted and the exhibits are labeled.   ¶ON PREMISES: snack bar, gift shop, exhibits of firefighting, trains, prehistory, Egypt, American Indians, toys and dolls   ¶NEARBY: all facilities   ¶NEARBY ATTRACTIONS: Since this museum is located in the heart of a city, all facilities are available except camping.

WYANDOTTE CAVE

Mailing address: Indiana Department of Natural Resources, Division of Forestry, Wyandotte, IN 47137   *Phone:* (812) 738-2782   *Directions:* On U.S. 460 and State Highway 62 about 35 miles west of Louisville, Kentucky.

Wyandotte Cave is one of the most celebrated caves in the United States. There are actually two caves open to the public: Big Wyandotte Cave and Little Wyandotte Cave. They were purchased by the Indiana Department of Natural Resources, together with 1100 acres of forest, from the Rothrock family in late 1966. This land will be added to the Harrison-Crawford State Forest which eventually will total 25,000 acres.

The early history of Big Wyandotte Cave includes visitation by prehistoric Indians. During the War of 1812 it was mined for saltpeter used in the manufacture of gunpowder. In 1820 Henry P. Rothrock, a pioneer from Pennsylvania, bought 4000 acres of this heavily forested land from the federal government for $1.25 per acre. He did not use the cave, however, but built a saw mill on the Blue River about ½ mile from the cave entrance. In 1850 explorers found a new entrance to Big Wyandotte, and their accounts of fine stalactites, stalagmites, helictites, gypsum flowers, and calcium and epsom salt crystals began to attract visitors from all over the world. Since that time visitors have enjoyed the unusual cave features which include Rothrock's Cathedral, a room 185 feet high, 360 feet long and 140 feet wide; an underground mountain 135 feet high; and a stalagmite in the so-called Senate Chamber which is 25 to 35 feet high and 71 feet in circumference. The present cave tour is 1½ to 2 hours in duration.

Little Wyandotte Cave is located 700 feet to the south of Big Wyandotte. Although much smaller it offers a very attractive trip for visitors who do not have the time or energy for a longer trip in the big cave. It is illuminated with fluorescent, photoflood and mercury lighting—a convenience for photographers. The tour through this cave takes approximately 45 minutes.

Two other tours are offered, both of them through Big Wyandotte Cave.

Five hour tour: This tour is by lantern light and is described as being available for the "vigorous, healthy, energetic cave enthusiast." The visitor crawls down to the bottom of the "Animal Pit" near the entrance and continues to an area mined for saltpeter. The trail leads up and down a series of hills, through several passages that require crawling on hands and knees, until the Senate Chamber is reached. The formation shown below is of the Pillar of Constitution seen on

Wyandotte Cave, Indiana. Photo by George Jackson.

this tour. The floor in this area is covered with ashes from fires the Indians built for ceremonies here.

Eight-hour tour: This tour is also by lantern light and is not as rugged as the Five-hour tour. It is primarily a walking tour and includes a few muddy crawl-ways. Visitors on this trip see most of what is seen on the two-hour regular tour, but this includes areas of maze-type passage and Milroy's Temple, one of the large rooms of the cave hot visited on the other trips.

Both of these longer tours are spelunking trips. While not hazardous, they do require endurance. Arrangements for them should be made by contacting: Property Manager, Wyandotte Cave, Route 1, Leavenworth, IN 47137.

¶OPEN: all year 8:00 A.M. to 5:00 P.M.   ¶GUIDED TOUR: Little Wyandotte: 45 minutes; Big Wyandotte: 1½ to 2 hours. Five and eight-hour trips by prior arrangement   ¶ON PREMISES: snack bar, gift shop, picnicking   ¶NEARBY: camping, motel, hotel, trailer camp   ¶NEARBY ATTRACTIONS: Corydon (first capitol of Indiana), Marengo Cave, Squire Boone Caverns, Spring Mill State Park, Bluespring Caverns, French Lick Mineral Springs, Cave River Valley Park.

IOWA

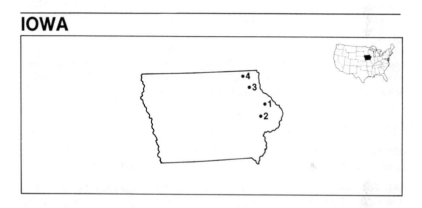

1 Crystal Lake Cave 2 Maquoketa Caves State Park 3 Spook Cave 4 Wonder Cave

THE GREAT PLAINS begin at the Mississippi River. For 800 miles the land rises gradually — about 7 feet per mile — to the mile-high foothills of the Rocky Mountains. The fertile region of Iowa has some of the finest farming soil in the United States, and early settlers who crossed the plains found natural grasses that reached heights of 14 feet.

The show caves of Iowa are limited to the northeast corner of the state which was missed by the glaciers that covered most of the area within the past 10,000 years. Most caves were prospected by lead miners in the middle of the nineteenth century.

CRYSTAL LAKE CAVE

Mailing address: R.R. 3, Dubuque, IO 52001   *Phone:* (319) 556-6451   *Directions:* About 5 miles south of Dubuque on U.S. Highway 52.

In 1880 a group of lead miners drilled 40 feet to find traces of a rich vein. In the process of mining for this deposit, they found very little lead, but instead discovered a seemingly unimportant natural cave. Many years later, in 1932, Bernard Markus, one of the original lead miners who discovered the cave, opened it to the public and named it Crystal Lake Cave, and it proved to be one of the largest in Iowa.

The tour covers 3/4 of a mile of passage — about 1/5 of the known corridors. The pattern follows a classic "city block" maze, each intersection at right angles to the main joint. The corridors are of uniform dimension with a level floor and almost straight passageways. Dripstone deposits are numerous and there are helictites, as well as aragonite crystals. The tour begins at one opening and exits through another; the only stairs are at either end.

¶OPEN: daily Memorial Day to Labor Day, 8:00 A.M. to 6:00 P.M.; May, September, and October, weekends only, 9:00 A.M. to 5:00 P.M.   ¶GUIDED TOUR: 35 minutes   ¶ON PREMISES: snack bar, gift shop, picnicking   ¶NEARBY: all facilities in Dubuque (5 miles)   ¶NEARBY ATTRACTIONS: Maquoketa Cave and State Park, St. Donatus (French village), Bellevue State Park, Wapsipinicon State Park, Clarke College, University of Dubuque, Mississippi Palisades State Park and caves (Illinois), Badger Mine (Wisconsin).

MAQUOKETA CAVES

Mailing address: R.R. 2, Maquoketa, IO 52060   *Phone:* (319) 676-3401   *Directions:* State Route 130, about 6 miles northwest of Maquoketa.

This 300-acre park contains 14 caves which can be reached via the foot trails that radiate from the parking area and picnic grounds. None of the caves are very large, and most are simply shelters — enlarged openings in the vertical limestone cliffs of a ravine. All of the caves were known to the Indians who hunted and roamed this area; evidence of their presence has been found in excavations in some of the cave floors. At the entrance of the park, a leaflet is available with a map showing locations of the caves. You may wish to choose from the following: Wide Mouth Cave, Hernando's Hideaway Cave, Up-N-Down Cave, Dance Hall Cave, Match Cave, Shinbone Cave, and a half dozen more. If you are traveling from out-of-state and do not know how to pronounce Maquoketa, the local people will help you with *Ma-KOH-key-tah*.

¶OPEN: daily, all year. Daylight hours   ¶SELF-GUIDED TOUR: bring lights, as some caves are not lighted   ¶ON PREMISES: snack bar, gift shop, camping, trailer camp, picnicking   ¶NEARBY: all facilities in the city of Maquoketa (6 miles)   ¶NEARBY ATTRACTIONS: Crystal Lake Cave, Bellevue State Park, Wapsipinicon State Park, St. Donatus (French village), Clarke College, University of Dubuque, Mississippi Palisades State Park and caves (Illinois), Badger Mine (Wisconsin).

SPOOK CAVE

Mailing address: McGregor, IO 52157   *Phone:* (319) 873-2144   *Directions:*
Off U.S. Highways 18 and 52, about 7 miles west of McGregor.

One of the springs supplying Bloody Run stream near Beulah Falls in the
northwest corner of Iowa was known to the early settlers because of the inter-
mittent noises that came from a tiny opening in the limestone rock. Many
people knew it was there, and some wondered about it; but Gerald Mielke decided to find out just what it was. He proceeded to blast out an opening, and
found an underground river leading back under the mountain. Construction
of a lock and dam outside controlled the level of the water, and a landing was
built over a cavern pool to take passengers inside. Aluminum boats, powered
with electric, front-end motors, now provide a quiet, effortless tour. There are
no longer any mysterious noises in the cave, and some say Mr. Mielke released
the spirits with his development. Since the time the cave was discovered in
1953, there have been many improvements in the surface area.

¶OPEN: daily Memorial Day through Labor Day. 8:30 A.M. to 5:00 P.M.
September and October, 9:00 A.M. to 4:00 P.M.   ¶GUIDED BOAT RIDE: 40
minutes   ¶ON PREMISES: snack bar, gift shop, camping, trailer camp, picnick-
ing, Spook Cave Water Wheel   ¶NEARBY: all facilities in private park and
resort area including swimming and fishing   ¶NEARBY ATTRACTIONS: Wonder
Cave, Kickapoo Caverns (Wisconsin), Moody's Museum, Echo Valley State
Park, Pikes Peak State Park, Villa Louis (Wisconsin), Effigy Mounds
National Monument.

WONDER CAVE

Mailing address: R.R. 2, Decorah, IO 52101   *Phone:* (319) 382-8871   *Directions:* 3 miles northeast of Decorah on Locust Road.

The northeast corner of Iowa is one of the most fertile sections of the country.
Black topsoil, nearly a foot deep, provides lush vegetation and excellent crops.
This particular area was also spared the scouring of the last glacier, and caves
developed here as the water drained off the surrounding ice pack.
   Wonder Cave, formed as glacial melt waters flowed through the rich soil,
provides a view of an excellent dome pit 150 feet high that received this highly
aggressive water from the surface and carved out a silo-like chamber in the
limestone. Later, dripping of water began to fill up the opening with flowstone
and draperies. The tour descends partway down the dome pit, and also follows
the joint that provided the original drainage pattern for the cave, which is
active and growing.

¶OPEN: daily Memorial Day to Labor Day 9:00 A.M. to 6:00 P.M.; weekends in

September and October, 10:00 A.M. to 6:00 P.M. ¶GUIDED TOUR: 40 minutes ¶ON PREMISES: gift shop, picnicking ¶NEARBY: all facilities in Decorah, 3 miles away ¶NEARBY ATTRACTIONS: Spook Cave, Echo Valley State Park, Effigy Mounds National Monument, Mystery Cave (Minnesota), Minnesota Caverns (Minnesota), Bily Clocks, Niagara Cave (Minnesota), Beaver Creek Valley (Minnesota).

KANSAS

LOCATED IN THE HEART OF THE Great Plains, Kansas has no show caves. A few small caves exist near the southeast border, but none have been opened to the public.

KENTUCKY

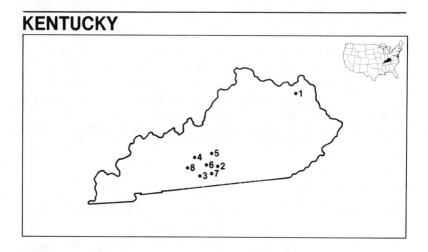

1 Carter Caves State Park 2 Crystal Onyx Cave 3 Diamond Caverns 4 Mammoth Cave National Park 5 Mammoth Onyx Cave 6 Old Original Cave 7 Onyx Cave 8 Park Mammoth Resort

THE CUMBERLAND PLATEAU of Kentucky and Tennessee is a fine example of the topographic feature called karst which sometimes develops in limestone terrains. Karst topography includes steep slopes with little level ground, many sinkhole depressions, rivers in deep gorges, disappearing streams, dry valleys, and springs. Beneath everything is a network of caves, often with streams flowing through them.

After the limestone beds were deposited some 300 million years ago in the

Mammoth Cave region, a sandstone deposit was laid down on top. When the land was uplifted and the water drained away after formation of the caves, this sandstone cap protected the limestone below from erosion. In places in Mammoth Cave, corridors in the limestone extend for miles below a flat sandstone ceiling. The caverns are mazes of passages that total hundreds of miles. Only a small number of the caves of Kentucky have been developed for visitors and these are clustered around the remarkable Mammoth Cave region.

CARTER CAVES STATE PARK

Mailing address: Olive Hill, KY 41164   *Phone:* (606) 286-4411   *Directions:* Take Exit 161 off Interstate 64, about 2 miles east on U.S. 60 to State Highway 182. Park entrance 8 miles north of Olive Hill.

The park is a 1000-acre site with excellent family recreational facilities. In addition to its varied programs, services and accommodations, the park contains a number of caves, although many are suitable mostly for equipped spelunkers. However, there are six caves listed for visitor tours. Only the first three of these caves are lighted; the others must be entered with portable lights.

Cascade Cavern Tour: This is the most scenic tour, and the cave has generally broad, high ceilings and the trip is fairly leisurely. Route includes total of about 100 stairsteps. Walking distance is 3/4 mile; outstanding features include a Counterfeiter's Room, Lake Room, Cathedral Room, "Hanging Gardens of King Solomon," and colorful formations.

Saltpeter Cave Tour: This is the historical trip and features saltpeter mining operations, Indian graves, old wall inscriptions and local historical information. Walking distance about 1 mile; duration about 45 minutes. Route includes about 30 stairsteps and a few places where the ceiling is slightly below head height.

X Cave Tour: This tour features cave formations, Lover's Leap, the Giant Stalactite, the "X," and dome pits. Walking distance is about 1/4 mile and the route includes about 60 stairsteps and narrow, winding passages.

Saltpeter Cave Spelunker Tour: (Unlighted trip by reservation only at Cave Information Desk in the Trading Post.) This is a rugged, strenuous trip through unlighted passages and crawlways. Participants must be in excellent physical condition, wear old clothes and bring a flashlight.

Bat Cave Tour: (Unlighted trip by reservation only at Cave Information Desk in the Trading Post.) This rugged and strenuous trip includes sections where extended stooping and squatting are required and also sections where there is difficulty avoiding water. Participants should be in good physical condition, wear old clothes and shoes, and bring a flashlight or lantern.

Laurel Cave/Horn Hollow Cave Tour: (Unlighted trip by reservation only at Cave Information Desk in the Trading Post.) This trip takes visitors to two caves. Walking distance is 1½ miles overland and ½ mile total in the caves; duration is about 2 hours. Route includes wading in water up to 15 inches deep, and some stooping is required. Participants should wear old clothes and shoes and bring a flashlight.

Young children are not permitted on the latter three trips. If visitors would like to visit any of the other caves in the park, it is necessary to obtain permission at the Cave Information Desk at the Trading Post.

¶OPEN: daily all year ¶GUIDED TOUR: Cascade Cavern Tour: 9:15, 10:00, and 11:00 A.M. and 1:00, 3:00 and 4:00 P.M.; Saltpeter Cave Tour: 9:15, 10:00, 11:00, 12:00 A.M. and 1:00, 2:00, 3:00 and 4:00 P.M. Saturday and Sunday 5:00 and 6:00 P.M.; X Cave Tour: 9:30, 10:30, 11:30 A.M. and 12:30, 1:30, 2:30, 3:30, 4:30 and 5:30 P.M.; Saltpeter Cave Spelunker Tour: Monday through Friday at 4:00 P.M.; Bat Cave Tour: 2:00 P.M. Tuesdays and Wednesdays and 10:00 A.M. Saturdays and Sundays; Laurel Cave/Horn Hollow Cave Tour: 2:00 P.M. Thursdays ¶ON PREMISES: restaurant, snack bar, gift shop, camping, motel, cottages, trailer camp, picnicking, golf, tennis, hiking, swimming, recreation room, nature films, boating, fishing, horseback riding, Nature Center, suffle-oard ¶NEARBY: scattered facilities ¶NEARBY ATTRACTIONS: Greenbo State Park, Grayson Lake State Park.

CRYSTAL ONYX CAVE

Mailing address: Cave City, KY 42127   *Phone:* (502) 773-2259   *Directions:* Off Interstate 65 to U.S. 31W, 2 miles south to Cave City.

In 1960 Kentucky cave explorer Cleon Turner set out to find a cave. In the tradition of earlier cave prospectors such as Floyd Collins, Turner used his intuitive sense, his years of experience in the Kentucky hill country, and a good deal of backbreaking effort. He also used what he called "cave sight." He said he could see a cave in the side of a mountain. One of those was a well decorated cave, with several levels and a most remarkable deposit of human bones esti-mated to be 4000 years old. The entrance used by prehistoric people has been lost, perhaps from a landslide or collapse of the cliff above. It is now only possible to visit the "burial grounds" by descending a stairwell from the upper cave first entered by Turner. From a vantage point on Pruitt's Knob visitors can look across the "barrens" of Kentucky, an area devoid of surface streams and named by the early settlers as they plodded their way west.

¶OPEN: daily all year. 9:00 A.M. to 6:00 P.M.   ¶GUIDED TOUR: 1 hour   ¶ON PREMISES: snack bar, gift shop, camping, picnicking, trailer camp   ¶NEARBY: restaurant, motel, gasoline   ¶NEARBY ATTRACTIONS: Mammoth Cave Na-tional Park, Mammoth Onyx Cave, Diamond Caverns, Barren River Lake, Western Kentucky University.

Crystal Onyx Cave, Kentucky.

DIAMOND CAVERNS

Mailing address: Park City, KY 42160   *Phone:* (502) 749-2891   *Directions:*
Exit 48 on Interstate 65, on State Highway 255, 1½ miles from Park City.

At the time Diamond Caverns was discovered in 1859, Mammoth Cave, only
five miles away, was already an internationally known attraction. The success
of Mammoth Cave caused every farmer who owned land in the area to seek out
a similar cave on his property and open it to the stream of people attracted to
Mammoth. At one time there were 17 caves vying for tourist business. Only a
few survived the "cave war" that developed. Location, access, and financing
proved to be more important to the success of the cave than the natural features it might have. Diamond Cave, with its large rooms, excellent location on
the main road, and good management, survived this period. Entrance is from
the lobby of Colonial Lodge and down a masonry stairway to the cave's main
passage. An easy trail, large formations, and dry constant temperature make it
an enjoyable experience.

¶OPEN: March to November, daily. 8:00 A.M. to 8:00 P.M.   ¶GUIDED TOUR: 1
hour   ¶ON PREMISES: restaurant, motel, gift shop, picnicking   ¶NEARBY:
camping, trailer camp, golf course   ¶NEARBY ATTRACTIONS: Mammoth Cave,
Mammoth Onyx Cave, Crystal Onyx Cave, scenic boat trips in Mammoth
Cave National Park on the Green River, Western Kentucky University.

MAMMOTH CAVE NATIONAL PARK

Mailing address: Mammoth Cave, KY 42259   *Phone:* (502) 758-2328   *Directions:* From Park City take State Highway 255 to park entrance; from there it is
5 miles to park headquarters. From Cave City take State Highway 70 ten miles
to park headquarters.

Mammoth Cave, the stellar attraction of this national park, is part of the longest continuous cave system in the world. More than 200 miles of mapped
passageway are beneath the 50,000-acre park, some of the cave having recently
been joined by exploration. With nearly 100 known caves on park property,
this region may be speleologically unique in the world.

The visitor to the National Park has a selection of trips available, and as he
drives into the park he may dial 1610 on his car radio for information about
these and other facilities. There are trips for people of all abilities, ranging from
special wheelchair tours to wild cave trips for groups. Below is a brief description of the tours. More detailed information may be found in leaflets at the
entrance gate.

Historic Tour: A two-mile, two-hour tour featuring: War of 1812 mining
operation, Indian artifacts, Fat Man's Misery, Mammoth Dome, and cave life.
At certain times this is a self-guided tour, but when it is guided it includes a

Mammoth Cave National Park, Inc. Photo by W. Ray Scott.

demonstration of torch throwing, the old method of lighting the cave. Restrooms are available. Daily 9:00 A.M. and 2:00 P.M. Tour limited to 160 people.

Frozen Niagara Tour: Features an impressive assortment of cave formations —stalactites, stalagmites, flowstone and draperies. This 1½ hour tour is the least strenuous offered, requiring a half-mile walk (1/4-mile into the cave, then retrace route). Daily 9:00 A.M., 11:00 A.M., 1:15 P.M., and 3:30 P.M. Tour limited to 120 people.

Half-day Tour (scenic): Features narrow winding cave passages, dome pits, breakdown, gypsum flowers and cave formations. This strenuous tour requires a four-mile, 4½-hour walk, including 700 steps, steep hills, and low passages. Food is available (cost not included in tour fee) in the Snowball Dining Room one hour after the tour starts. The Half-day Tour includes the Frozen Niagara Tour. Restrooms are available underground. Trip begins daily at 12:00 noon. Tour limited to 160 people.

Lantern Tour: A nostalgic tour by lantern light featuring: The natural cave entrance, War of 1812 mining operation, Mummy ledge, tuberculosis huts, large cave rooms, Indian artifacts and torch-throwing demonstration. Includes half mile of Historic Tour route. Walking distance about three miles. Tour duration about three hours. Restrooms not available. Lanterns provided. Summer only, April to October; check time schedule.

Great Onyx Tour: Features an assortment of speleothems in a cavern not connected with Mammoth Cave. Tour includes a three-mile bus ride though a hardwood forest with interpretation of surface features by guides. Walking distance about one mile, which includes a quarter-mile walk from the bus to the cave entrance. Tour duration about 2½ hours. Restrooms available. Lanterns provided. Summer only, April to October; check time schedule.

Wild Cave Tour (reservations only): A spelunking adventure limited to 14 people. Walking distance about five miles; tour duration is six hours. Rugged, strenuous tour over unimproved trails and through crawlways. Participants must be in excellent physical condition. Wear old clothes, sturdy shoes, and gloves. Hard hats, headlamps and knee pads provided. Regulations for Wild Cave Tour are as follows:

1. Individuals may make reservations by phone or in person only Monday through Friday, 8:00 A.M. to 4:30 P.M. (Central Time Zone). For reservations call (502) 758-2328 and ask for Wild Tour reservation clerk.

2. Participants must be 16 or over, show proof of age (birth certificate or valid driver's license), and must purchase tickets no later than 15 minutes before departure time.

3. Tickets must be purchased in person only on the day of the tour.

Wild Cave Tour leaves at 9:30 A.M. daily in summer; weekends only, November to March.

Wheelchair Tour: Restricted to wheelchair patients and their assistants. Tour distance about a half-mile. Duration about 1½ hours. Daily 9:00 A.M. Handicapped Groups by reservation (consisting of five or more wheelchair viisitors).

As in all National Parks there are other activities available for the public. Schedule of guided walks is posted and includes: Green River Bluffs, River Styx, Cedar Sink, Beaver Pond, and evening programs at campfire circle in the campground.

¶OPEN: Cave tours as stated above. Check with Visitors' Center for current time. Visitors' Center open 8:00 A.M. to 5:00 P.M. (Central Time Zone). ¶OPEN: restaurant, snack bar, gift shop, photo shop, camping, motel, hotel, underground restaurant, trailer camp, cabins, lodge, cottages, picnicking, nature trails ¶NEARBY: all facilities ¶NEARBY ATTRACTIONS: Scenic boat trips on the Green River, Mammoth Onyx Cave, Diamond Caverns, Crystal Onyx Cave, Abraham Lincoln National Historical Site, resort area.

MAMMOTH ONYX CAVE

Mailing address: Box 527, Horse Cave, KY 42749   *Phone:* (502) 786-2634
Directions: At Horse Cave, Exit 58 on Interstate 65, and State Highway 335, three miles west of Horse Cave.

Over a century and half ago, young Martha Woodson, so legend says, discovered Mammoth Onyx Cave. The year was 1799 and this area of Kentucky was part of the "roaring wilderness" which attracted settlers from the eastern seaboard. It was not until 1921 that the cave was developed for visitors providing an easy, near-level tour of highly decorated rooms. A stream of clear green water still runs through the main passage, and visitors sometimes are

Mammoth Onyx Cave, Kentucky.

able to see the blind fish, salamanders or white crayfish that still inhabit the cave pools. Equipped with electric lights in 1922, it was one of the first caves in the area to be so lighted. The system has since been updated with indirect lighting which brings out the subtle, natural colors of the underground architecture.

¶OPEN: all year. Summer 8:00 A.M. to 6:00 P.M.; winter 8:00 A.M. to 4:00 P.M. ¶GUIDED TOUR: 1 hour   ¶ON PREMISES: gift shop, picnicking, wildlife preserve

¶NEARBY: motel, restaurant, trailer park, camping   ¶NEARBY ATTRACTIONS: Mammoth Cave National Park, Abraham Lincoln Birthplace National Historical Site, Diamond Caverns, Crystal Onyx Cave, 100 Domes Cave, Jesse James Cave, Onyx Cave, Old Original Cave.

OLD ORIGINAL CAVE

Mailing address: Route 1, Box 149, Wondering Woods, KY 42160   *Phone:* (502) 749-5521   *Directions:* On State Highway 70 at intersection of State Highway 255, adjacent to the main entrance to Mammoth Cave National Park.

The site is a 1900s community which includes a museum, farm, blacksmith's shop, restaurant and demonstrations by artists and craftsmen. In the Village of Wondering Woods is the entrance of Old Original Cave. The cavern is highly decorated, and following the tour, visitors are taken on a guided nature walk through the Majestic Forest which surrounds it.

¶OPEN: daily from Memorial Day to Labor Day 9:00 A.M. to 4:00 P.M. Weekends only, in May and September   ¶GUIDED TOUR: Tours depart every hour and are 45 minutes in duration. Part of the tour is lighted by lanterns.   ¶ON PREMISES: restaurant, snack bar, gift shop, picnicking, wax museum, horse-drawn carriage rides, artisans' shops, farm   ¶NEARBY: camping, motel, trailer camp   ¶NEARBY ATTRACTIONS: Mammoth Cave National Park, Crystal Onyx Cave, Diamond Caverns, Mammoth Onyx Cave, Park Mammoth Resort, Barren River Lake.

Old Original Cave, Kentucky.

ONYX CAVE

Mailing address: Cave City, KY 42127   *Phone:* (502) 773-2375   *Directions:* Take Cave City Exit 48 on Interstate 65, then 1 mile West on State Highway 70.

The entrance to this cave is within sight of Interstate Highway 65. This is not a coincidence, for in 1972 Kentucky explorer Cleon Turner was asked to survey a piece of property here for the potential of having a cave suitable for exhibition. Mr. Turner, who immodestly claims to have abilities to see caves through the soil, is actually a skillful geomorphologist who knows the Kentucky cave country and the indications that predict the presence of caves. He selected a spot to dig, which resulted in the discovery of this small cave possessing three dome pits and fine formations. The cave was equipped with electric lights and trails and was opened to the public in 1973. The developers have added a number of attractions to the surface area including a water slide, chair lift and frontier village.

¶OPEN: all year. Summer 8:00 A.M. to 8:00 P.M.; winter 8:00 A.M. to 4:00 P.M. ¶GUIDED TOUR: 30 minutes   ¶ON PREMISES: snack bar, gift shop, picnicking ¶NEARBY: motel, restaurant, all facilities in Cave City   ¶NEARBY ATTRACTIONS: Mammoth Cave National Park, Diamond Caverns, Crystal Onyx Cave, Mammoth Onyx Cave, Park Mammoth Resort, Old Original Cave, Barren River Lake State Park, Abraham Lincoln Birthplace National Historic Site, Western Kentucky University.

PARK MAMMOTH RESORT

Mailing address: Park City, KY 42160   *Phone:* (502) 749-4101   *Directions:* Exit at Park City off Interstate 65, take U.S. 31W south 1 mile.

This resort is located on 2000 acres of wooded land and provides a variety of recreational and convention facilities for visitors. Among them are Jesse James Cave and Hundred Dome Cave. Hundred Dome Cave has been operated more than 100 years. Slave Cave, also on the resort land, was used as a way station for escaping slaves and features a perfectly preserved lime rock cooking stove, purported to have been made by a fugitive who hid there for more than three years.

¶OPEN: all year   ¶GUIDED TOUR: Hundred Dome Cave: 8:00 P.M. May to mid-September. Two-hour trip which includes train ride. Jesse James Cave: 9:00 A.M., 10:30 A.M., 12:30 P.M. and 2:00 P.M. year round. Tour length 1½ hours ¶ON PREMISES: restaurant, snack bar, gift shop, motel, picnicking, indoor pool, 3½-mile miniature train ride, horseback riding, tennis, sauna bath, golf, convention facilities (up to 400 persons)   ¶NEARBY ATTRACTIONS: Mammoth Cave National Park, Crystal Onyx Cave, Diamond Caverns, Mammoth Onyx Cave, Onyx Cave, Old Original Cave, Barren River Lake.

LOUISIANA

THE DELTA OF THE Mississippi River dominates the landscape of Louisiana. There are no show caves in the state and only a few wild caves along the Texas border in the "hill country."

MAINE

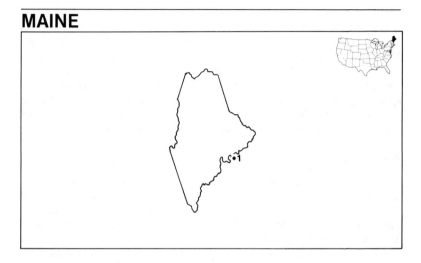

1 Anemone Cave

IN THE LONG GEOLOGICAL HISTORY of what we now know as the state of Maine, there have been numerous periods of uplift and subsidence. The most recent occurrence (in geologic terms) is the flooding of the coastal area as the glaciers melted forming today's irregular rocky coastline. An airline distance of 228 miles, the coast actually meanders 3478 miles from the Canadian border to the New Hampshire line. It is subject to the direct pounding of the Atlantic Ocean and contains thousands of erosional features including caves, arches, blowholes, and tidal basins. The only show cave in the state can be seen at Acadia National Park, but there are many other caves on private property and some on state lands.

ANEMONE CAVE

Mailing address: Acadia National Park, Bar Harbor, ME 04609 *Phone:* (207) 228-3338 *Directions:* Reached from U.S. Route 1 at Ellsworth onto State Route 3 to Bar Harbor. From Bar Harbor ¾ mile south on Ocean Drive to Anemone Cave parking area.

Anemone Cave is one of the finest examples of a sea cave. Located on the eastern shoreline of Mount Desert Island in Acadia National Park, it has been carved by sea action for a depth of 82 feet. Below the overhanging roof of the cave is a tidepool containing a great wealth of color and a variety of marine creatures. Conspicuous among these are the sea anemones for which the cave is named, as well also as sea urchins, yellow and green sponges, starfish, brittle stars, green crabs, limpets, barnacles, periwinkles, purple rock snails, and sea cucumbers.

Visitors should note that the cave is one fifth of a mile from the parking area along a well-marked trail. The cave is best visited in the morning before 10:00 while the sunlight illuminates the cave's interior. Favorable tide conditions combined with the morning sun offer the best viewing. The most advantageous tide is half tide or lower; at this time one may enter the cave dryshod. The approach to the cave is over rough, rocky terrain that requires care and appropriate footwear.

¶OPEN: May 1 to December 1 all day   ¶SELF-GUIDED TOUR: no lights needed ¶ON PREMISES: camping, trailer camp, picnicking   ¶NEARBY: restaurant, snack bar, gift shop, motels, hotels, cabins, and all facilities in Bar Harbor ¶NEARBY ATTRACTIONS: Acadia National Park, Moose Point State Park, Camden Hills State Park, Lamoin State Park, Maine coast shore resorts.

MARYLAND

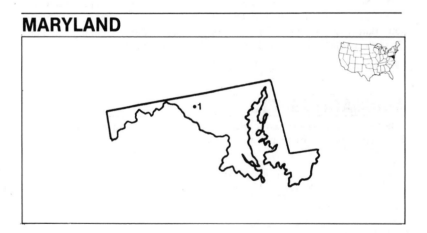

1 Crystal Grottoes Caverns

MARYLAND CONTAINS features of the Atlantic Coastal Plain, the Piedmont Plateau, and the Appalachian Mountains, but there are caves only in the narrow mountain corridor in the middle of the state. This region, on the eastern slope of the mountain range, contains limestone beds with some caves noted for fossil remains of extinct animals. One cave is open to the public as a visitor attraction.

CRYSTAL GROTTOES CAVERNS

Mailing address: Route 1, Boonsboro, MD 21713   *Phone:* (301) 432-6336
Directions: On State Route 34, about 1 mile south of U.S. Alternate 40.

Crystal Grottoes Caverns is located only a few thousand feet from the Antietam Battlefield, scene of one of the bloodiest conflicts of the Civil War. Today it is a peaceful spot with a quiet stream, picnic grove, and a spring that flows from a rocky bluff into a rocky creek. No one suspected the presence of a cave until a quarry operation in the rock face broke into it in 1920. Exploration of the passageways convinced quarrymen that they should give up the mining operation and open the cave to the public. It was opened in 1922 and has remained in the same family ever since. A small house and ticket office contains a stairway which leads down into the cave's entrance room. In the summer the door above is left open and the cave's natural air conditioning keeps the house cool. The cave follows a joint pattern—several parallel passages that are interconnected at near-right angles. There are fine formations and interesting wall and ceiling solution channels. The cave exit (outside the little house) is nearly at the level of the outside stream.

¶OPEN: all year, Summer 9:00 A.M. to 6:00 P.M.; winter, 11:00 A.M. to 5:00 P.M. ¶GUIDED TOUR: 1 hour ¶ON PREMISES: gift shop, picnicking ¶NEARBY: restaurant, motel, hotel, camping ¶NEARBY ATTRACTIONS: Antietam Battlefield, Harpers Ferry National Historic Park, Cunningham Falls State Park, Gambrill's State Park, Catoctin Mountain Park.

MASSACHUSETTS

SHELTER CAVES, marble caves, and boulder caves are found in Massachusetts, but there are no caves open to the public.

MICHIGAN

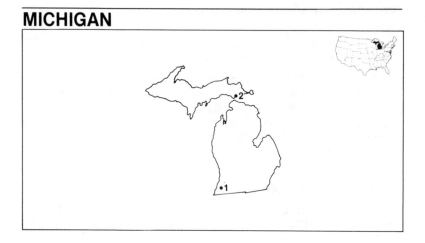

1 Bear Cave 2 Mackinac Island Caves

DURING THE GREAT ICE ADVANCE the waters that flowed north to Hudson Bay were trapped in a huge lake covering an area larger than all the present Great Lakes. As the ice melted about 10,000 years ago the water drained away, leaving the present lakes. The land that emerged includes all of the state of Michigan. The basement rock of this area is billions of years old and has been worn down by erosion and covered by sedimentary rocks. The mature, nearly level landscape is heavily wooded in the northern part of the state. There is an interesting limestone area at the northern end of the main peninsula, but there are no caves of great size. Two areas contain small caves open to the public.

BEAR CAVE

Mailing address: Route 1, Box 550, Buchanan, MI 49107 *Phone:* (616) 695-3050 *Directions:* 4 miles north of Buchanan on Red Bud Trail.

The area of the United States which is now the state of Michigan was covered with glaciers at least four times in the recent geologic past. The glacial ice pack crushed or filled most of the caves that might have existed in the limestone beds of the area. In the lower part of the state, the limestone was covered with 200 to 500 feet of glacial debris, preventing the development of caves. In Bear Cave there appears to be an unusual condition which is unique among the show caves of this country. The limestone that provided the calcite for this cave is buried deep below it. At some time within the past 50,000 years, a spring in the limestone carried minerals from this bed up to the surface and deposited a layer of secondary material on top of the ground. This might have been a hot spring, or at least a mineral spring, as is seen today in certain parts of the country. This bed of calcareous material, called tufa, is porous and soft. After the glaciers left, the bed began to dissolve in the surface water that seeped through the soil above. Bear Cave, with 150 feet of passage, has developed in this material, and

111

the formations and walls show the porous nature of the rock that is formed by this method. It is an unusual cave of particular interest to geologists.

¶OPEN: daily, June, July and August, 10:00 A.M. to 5:00 P.M. May, September, and October, weekends only, 10:00 AM to 5:00 P.M. ¶GUIDED TOUR: 30 minutes ¶ON PREMISES: snack bar, figt shop, camping, trailer camp, picnicking, playground, boating, fishing ¶NEARBY: restaurant, motel, hotel ¶NEARBY ATTRACTIONS: Lake Michigan, Fort Saint Joseph Museum, Warren Dunes State Park.

MACKINAC ISLAND CAVES

Mailing address: Mackinac Island State Park, Mackinac Island, MI 49727 *Phone:* (906) 847-3212 *Directions:* The island is serviced by two passenger ferries from St. Ignace and Mackinaw City.

This small but historic island located in the narrow straits between Lake Michigan and Lake Huron had an important military location in the early days of the U.S. before the railroads displaced the rivers and lakes as main thoroughfares. When the island no longer had a military use, it was developed as a beautiful resort for summer visitors, and at the turn of the century it was the social center for the vacationing wealthy elite. Today it is just as beautiful, but is now the

Arch Rock, Mackinac Island State Park, Michigan.

popular destination of young people and families who enjoy the isolation and beauty of the place.

There are several small caves on the island that are of historic interest, although only of minor speleological importance. There are small shelters in the limestone rock with wonderfully romantic names such as Skull Cave, Cave of the Woods, Fairy Kitchen, Devils Kitchen, and Crack in the Island. Most of the caves can be reached by hiking on the trails that crisscross the island, but some like Skull Cave and Devils Kitchen are accessible by horse and carriage, a picturesque and colorful way to reach them. Arch Rock, a natural bridge on the northeast coast, is reached by a pleasant hike along Lake Shore Road which circles the island.

¶OPEN: all year. There are year-round residents but visitors come generally in summer only. Check with park headquarters regarding transportation and accommodations   ¶ON PREMISES: All facilities are available on the island. ¶NEARBY ATTRACTIONS: Fort Mackinac, Fort Holmes, Mackinac Island City, Burt Lake State Park, Wilderness State Park, Cheboygan State Park, Straits State Park, Foley Creek State Park, Michlimackinac State Park.

MINNESOTA

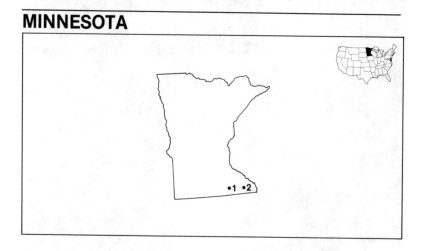

1 Mystery Cave/Minnesota Caverns 2 Niagara Cave

THERE ARE NEARLY 8000 lakes in Minnesota, remnants of a single ancient lake called Lake Agassiz that extended into Canada. The silt deposits of this lake cover ancient rocks perhaps two billion years old and have formed one of the most level regions in the world. The only area of the state with limestone exposed is in the southeast corner, near the Iowa border. This small section escaped the last two glaciations, and melting waters from the surrounding ice cut down into the limestone, making extensive caves.

MYSTERY CAVE and MINNESOTA CAVERNS

Mailing address: Rural Route 2, Spring Valley, MN 55975   *Phone:* (507) 937-3251   *Directions:* Off U.S. 16 and 63, about 6 miles southeast of Spring Valley.

Mystery Cave and Minnesota Caverns, Minnesota.

These two caves, both part of the Mystery Cave system, were discovered in 1937 and have been explored for more than 15 miles. Referred to as Mystery I and Mystery II, each has an independent tour.

Mystery I contains about one mile of passage available for the visitor. The cave is entered through the original entrance, at the disappearance of Root River. The water from this stream flows underground for 1½ miles, where it reappears in a spring. The cave is formed in two levels: the visitor traverses the upper level, which is dry; the lower level carries the stream and flood waters.

Mystery II entrance is approximately 1½ miles from the first entrance, and the cave contains corridors extending for more than a mile through the limestone.

Both caves show the action of solution, with convoluted walls and dissolved channels and pockets in the limestone. Because both caves are under the same management and take about an equal amount of time to visit, it is recommended that visitors allow sufficient time to see both.

¶OPEN: May through September, 10:00 A.M. to 4:00 P.M. daily   ¶GUIDED TOUR: 1 hour (for each cave tour)   ¶ON PREMISES: snack bar, gift shop, camping, picnicking, fishing, and hiking   ¶NEARBY: motel, restaurant, hotel, cabins and facilities in Spring Valley (6 miles)   ¶NEARBY ATTRACTIONS: Forestville State Park, Lake Louise State Park, Wonder Cave (Iowa), Niagara Cave.

NIAGARA CAVE

Mailing address: RDF Route #2, Harmony, MN 55939   *Phone:* (507) 886-6606   *Directions:* Off U.S. 52 on State Highway 139, about 4 miles southwest of Harmony.

The southwest corner of Minnesota was not covered with ice in the last glaciation and this permitted limestone caves here to reach imposing proportions. Niagara Cave was formed along the joints and cracks in the nearly level limestone, making a maze-like pattern of intersecting passages. Discovered in 1928 by a farmer who was searching for three lost pigs, the cave was developed in 1934 and has been exhibited ever since. A building was erected over the entrance and a stairway descends into the cave below. The walls show solutional sculpturing, typical of passages that have been flooded with water. The small stream that now courses through the corridors plunges off a precipice into a dome pit. The visitor can stand on a bridge over this chasm and look downward to the water dropping 60 feet to the bottom of the pit, and upward 70 feet to the top of the dome.

¶OPEN: Memorial Day to Labor Day, 9:00 A.M. to 5:00 P.M. daily   ¶GUIDED TOUR: 1 hour   ¶ON PREMISES: snack bar, gift shop, camping, trailer camp, picnicking   ¶NEARBY: motels and cabins   ¶NEARBY ATTRACTIONS: Lake Louise State Park, Mystery Cave, Wonder Cave (Iowa), Beaver Creek Valley State Park, Forestville State Park, Norwegian Museum (Iowa).

Niagara Cave, Minnesota. Photo by S. W. Lock.

MISSISSIPPI

MOST OF MISSISSIPPI is covered with the sands and gravels of the Gulf Coastal Plains. There are some wild caves in the northeast corner of the state where the Cumberland Plateau intrudes into the state but there are no show caves.

MISSOURI

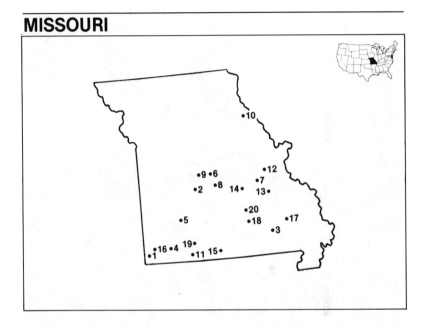

1 Bluff Dweller's Cave 2 Bridal Cave 3 Cave Spring Onyx Caverns 4 Crystal Caverns 5 Fantastic Caverns 6 Fantasy World Caverns 7 Fisher's Cave 8 Indian Burial Cave 9 Jacob's Cave 10 Mark Twain Cave 11 Marvel Cave 12 Meramec Caverns 13 Onondaga Cave 14 Ozark Caverns 15 Ozark Underground Laboratory 16 Ozark Wonder Cave 17 Rebel Cave 18 Round Spring Cave 19 Talking Rocks Cavern 20 The Sinks

THERE ARE THREE GENERAL regions of Missouri: the prairie and rolling plains of the north, the Ozark uplift of the southwest, and the flat delta area of the southeast. Most of the caves occur in the Ozark Mountains plateau area. While the crests of these mountains are relatively level, they are eroded into sharp ridges and deep valleys. Historically the caves of this region served as shelter and places of sanctuary for Indians. Settlers mined saltpeter and guano from the caves; outlaws, counterfeiters, and moonshiners occupied others. There are 20 caves open to the public and hundreds of others suitable for exhibition but not located conveniently for the traveling public. Missouri probably has more caves than any other state, and deserves the title "Cave State."

BLUFF DWELLER'S CAVE

Mailing address: Route 2, Box 229, Noel, MO 64854   *Phone:* (417) 475-3666
Directions: State Highway 59, about 2 miles south of Noel.

Bluff Dweller's Cave, Missouri.

The soft limestone cliffs of southwest Missouri have many projecting overhangs that were used by Indians for campsites. One of these shelters was explored in 1927 by some workers on a highway, where they discovered Indian artifacts in the rubble and the opening to a cave. Called Bluff Dweller's Cave, it proved to be a series of water-dissolved corridors following the joints of the limestone. The nearly level bedding is clearly visible throughout the cave, and the outline of the corridors takes on an "inkblot" design of equal patterns on both walls. There are excellent examples of fossils, and some of the strata show animals that helped form the limestone many millions of years ago. Most of the cave is dry, but a shallow lake near the exit has a fine example of a rimstone dam. Exit from the cave is through another shelter opening in the cliff. The tour through the cave has brought the visitor back within a few feet of the starting place.

¶OPEN: all year, 8:00 A.M. to 6:00 P.M.   ¶GUIDED TOUR: 45 minutes   ¶ON PREMISES: gift shop, picnicking   ¶NEARBY: snack bar, restaurant, motel, camping, hotel, cabins, trailer camp   ¶NEARBY ATTRACTIONS: Ozark Wonder Cave, Truitt's Cave, Crystal Caverns, Marvel Cave, Talking Rocks Cavern, Table Rock Lake resort area, Roaring River State Park.

BRIDAL CAVE

Mailing address: Camdenton, MO 65020   *Phone:* (314) 346-2676   *Directions:* Missouri Highway 5, about 3½ miles north of Camdenton.

Located on an arm of the Lake of the Ozarks, Bridal Cave has access by boat as well as by car. A fieldstone building over the entrance makes a meeting place for visitors, as well as the for the guests of the many wedding parties which come to the cave. More than 600 couples in recent years have taken their vows in the natural chapel within the cave. Legend has it that it all started with the marriage of the Osage Indian Princess, Irona, to the Indian brave, Prince Buffalo. The cave is notable also for the flowstone draperies that cover the entire wall of the main chamber.

¶OPEN: all year. Summer 9:00 A.M. to 7:00 P.M.; winter 9:00 A.M. to 5:00 P.M. ¶GUIDED TOUR: 40 minutes   ¶ON PREMISES: gift shop   ¶NEARBY: restaurant, snack bar, camping, motel, hotel, trailer camp, picnicking   ¶NEARBY ATTRACTIONS: Lake of the Ozarks resort area, Jacob's Cave, Ozark Caverns, Indian Burial Cave, Fantasy World Cave.

CAVE SPRING ONYX CAVERNS

Mailing address: Route 1, Van Buren, MO 63965   *Phone:* none listed   *Directions:* On U.S. 60, 1 mile west of Van Buren.

Located only a few hundred feet off the main highway, this small cave has a varied history, and is reported to have been used as a hiding place station on

Bridal Cave, Missouri.

the Underground Railroad — a means by which slaves were moved to freedom. The cave is also reputed to have been used by Jesse James as a hideout after the Gad's Hill Train Robbery.

Opened to the public in 1929, the cave and the lands above were operated as a nonprofit fishing and hunting club. In 1965 it was suggested that the cave was actually the original entrance to famous Big Spring (only a mile away), so the name was changed to Big Spring Onyx Caverns. Recently the cave changed ownership and the name has reverted to Cave Spring Onyx Caverns. The entrance is off the parking lot through an iron gate. After a few steps downward, visitors are in the main passage of a narrow winding corridor of the cave. The floor is a wooden boardwalk over a small stream that flows through the cave. It is difficult to imagine where anyone could hide in such a twisted crowded spot. There are almost no bare walls exposed; the flowstone and formations have nearly filled the passage. Bare light bulbs light the way and it is necessary to proceed single file, as the passageway is too narrow for two people to pass. Exit is from the deepest part of the tour, about 35 feet underground, where a masonry stairway leads to the surface.

¶OPEN: April 1 to October 1, daily   ¶SELF-GUIDED TOUR: There are electric lights, but flashlights are recommended.   ¶ON PREMISES: camping, trailer hookups, showers, picnicking   ¶NEARBY: motels, hotel, swimming, fishing   ¶NEARBY ATTRACTIONS: Big Spring, Fremont Tower Camp, Hawes Camp, Ozark National Scenic Riverways, Current River.

CRYSTAL CAVERNS

Mailing address: Route 3, Cassville, MO 65610   *Phone:* (417) 847-4238   *Directions:* On State Business Route 37, about ½ mile north of Cassville business district.

The Ozark area of southwest Missouri is a verdant region of rolling hills, forests and rivers. There are thousands of caves in the limestone, many of which have no natural entrances, their presence having been discovered by artificial

Crystal Caverns, Missouri.

means such as by cutting roads and by quarrying. Crystal Caverns was discovered by people who noticed a persistent winter "fog" that came from a crevice in the limestone. The smoke effect is caused by the condensation of warm, humid cave air as it escapes into the colder air outside. Excavation of this opening revealed a large collapsed dome spanning more than 150 feet. The jumble of breakdown beneath has been undisturbed for thousands of years, as formations caused by the percolating waters from above have cemented them together in many places. Although the cave is principally one large room, the interstices in the breakdown make it seem to be a series of rooms. There are fine examples of aragonite crystals and a display of crinoid stems in the Fossil Room at the end of the tour. Local legend states that an Indian maiden entered the cave through an underground connection to nearby Roaring River. There is no evidence to support this, but there are abundant ancient campsites and artifacts in the vicinity that show this was a popular area for early Indians.

¶OPEN: March to October daily 9:00 A.M. to dark   ¶GUIDED TOUR: 35 minutes   ¶ON PREMISES: snack bar, gift shop, picnicking   ¶NEARBY: motels, hotel, trailer camp, camping   ¶NEARBY ATTRACTIONS: Roaring River State Park, Table Rock Lake, Marvel CAve, Talking Rocks Cavern, Ozark Wonder Cave, Bluff Dwellers Cave, Onyx Cave (Arkansas), Cosmic Cavern (Arkansas).

FANTASTIC CAVERNS

Route 20, Box 1925, Springfield, MO 65803   *Phone:* (417) 833-2010   *Directions:* Off Interstate 44 on U.S. 13, 1½ miles north to Fantastic Caverns Road.

Temple Cave, Percy Cave, and Knox Cave were some of the local names previously given this 1400-foot tunnel in limestone. At one time it was owned by a local group of Ku Klux Klan members, and secret meetings were held in the cave auditorium until the property was lost by foreclosure. Today, a jeep pulling a small trailer equipped with seats for passengers provides a novel and pleasant tour of this large and highly decorated cave. One of the underground features is a group of columns that connect floor to ceiling providing an illusion of the vehicle passing through a grove of trees. The unusual method of traveling underground is perhaps the most remembered part of the tour.

¶OPEN: daily, all year. 8:00 A.M. to dusk.   ¶GUIDED TOUR: by jeep, 45 minutes   ¶ON PREMISES: restaurant, gift shop, picnicking   ¶NEARBY: all facilities in Springfield (3 miles)   ¶NEARBY ATTRACTIONS: Crystal Cave, Wilson's Creek Battlefield, Springfield, Marvel Cave, Talking Rocks Cavern.

FANTASY WORLD CAVERNS

Mailing address: Box 6, Osage Beach, MO 65065   *Phone:* (314) 392-3080   *Directions:* ¾-mile off U.S. 54 between Bagnell Dam and Eldon.

Fantastic Caverns, Missouri.

Fantasy World Caverns, Missouri.

The entrance of Fantasy World Caverns is in a small hollow of a densely wooded area. A natural flat arch spans more than 50 feet and the inviting passage has been a shelter for hibernating animals, Indians, and early settlers for thousands of years. Opened to the public in 1950 as Stark Caverns, it has been electrically lighted and the natural level floor only slightly modified to permit an easy tour. Noted for spongework ceilings and walls which provide interesting contrasts of light and shadow, the cave also has a complex floor plan. Several small streams converge and form a pool known as the "Fountain of Youth," shown above.

¶OPEN: April to September, daily. 9:00 A.M. to 5:00 P.M.   ¶GUIDED TOUR: 1 hour   ¶ON PREMISES: gift shop, picnicking   ¶NEARBY: restaurants, motels, camping, trailer camp   ¶NEARBY ATTRACTIONS: Lake of the Ozarks resort area, Jacob's Cave, Ozark Caverns, Bridal Cave, Bagnell Dam.

FISHER'S CAVE

Mailing address: Meramec State Park, Sullivan, MO 63080   *Phone:* (314) 468-3388   *Directions:* Exit 224 at Sullivan, take State Highway 155 east about 3 miles.

One of the many caves located in the bluffs of the Meramec River, Fisher's Cave has been a popular visiting spot for more than 100 years. First Indians, than local settlers entered and left evidence of their explorations. In 1910 it was opened as a private show cave and in 1933 it was acquired by the State and has been part of the park system ever since. The entrance is located near the level of the Meramec River and is a fine example of the horizontal tunnel-like entrances of Missouri. The corridor continues for 600 feet of nearly the same height and width. The only variance in the height is at the last half of the passage where it is necessary for the average person to stoop. The trip is worth the discomfort, for the main passage that this corridor leads to is broad and high and the rest of the tour requires no more effort than the surface hike to the cave. Six bridges have been built in the cave to span the small stream that meanders through the clay-floored main channel. The cave is not electrically lighted; lanterns supplied by the rangers provide interesting shadows and a sense of exploration and excitement. While there are more than 20 caves in Meramec Park, only Fisher's Cave is shown by guided tour. The others require special equipment and permission from park authorities.

¶OPEN: May through October, 9:00 A.M. to 5:00 P.M.   ¶GUIDED TOUR: 1 hour. Lantern tour, lights provided   ¶ON PREMISES: snack bar, gift shop, camping, picnicking, trailer camp   ¶NEARBY: all facilities   ¶NEARBY ATTRACTIONS: Onondaga Cave, Meramec Caverns, Meramec Spring State Park, Six Flags over Mid-America.

INDIAN BURIAL CAVE

Mailing address: P.O. Box 242, Osage Beach, MO 65065 *Phone:* (314) 348-2207 *Directions:* Located south of Bagnell Dam. Take County Route V off State Highway 54.

The limestone bluffs along the Osage River contain a number of caves which were used by Indians for shelter. One of these, near the Bagnell Dam, has a spectacular view of the Osage valley from an observation deck overlooking the river. The cave, located more than 100 feet above a parking area, is reached by means of a small inclined railway that takes visitors to a sheltered recess in the cliff where artifacts and relics of Indian burials are displayed. The actual, *in situ*, archaeological remains are behind glass and include the skeletons of reputed 1500-year-old burials. The cave trip is by means of a flat-bottomed boat that follows an underground stream. The cave route terminates at a waterfall flowing over natural rimstone pools at the rear of the cave.

¶OPEN: April to October, 9:00 A.M. to 5:00 P.M. daily ¶GUIDED TOUR: 30 minutes. Includes boat ride and inclined railway ¶ON PREMISES: gift shop ¶NEARBY: motel, restaurant, camping, trailer park, Osage Beach resort area ¶NEARBY ATTRACTIONS: Jacob's Cave, Fantasy World Caverns, Bridal Cave, Lake of the Ozarks State Park, Ozark Lake resort area.

JACOB'S CAVE

Mailing address: Route 2, Versailles, MO 65084 *Phone:* (314) 378-4374 *Directions:* Off State Highway 5 about 6 miles south of Versailles.

More than a hundred years ago, a lead and zinc prospector searching the bluffs along the Gravois Creek discovered this cave. He found no lead or zinc, but he did uncover an interesting and beautiful cavern containing prehistoric bones and fossil teeth. A building now covers the original entrance; paved trails permit an easy tour (even suitable for wheelchairs). A small stream flows throughout the main section of the cave, but disappears to a lower level before reaching the entrance. The stream did not form the cave, which appears to have formed below the water table when the creek valley did not exist. The walls show the sponge-work type of solution openings that are typical of caves formed beneath the water table. After the valley was formed (in fairly recent geologic time) the water drained out of the cave and the formations now covering the walls were deposited. The bone bed located near the entrance room was evidently the bottom of a natural trap, possibly a fissure that small animals fell into and were unable to escape from. Their remains over the centuries have formed a deposit here of paleological interest.

¶OPEN: all year. Summer, 9:00 A.M. to 5:30 P.M.; winter, 9:30 A.M. to 5:00 P.M.

¶GUIDED TOUR: 40 minutes   ¶ON PREMISES: snack bar, gift shop, picnicking, rock shop   ¶NEARBY: motel, restaurant, camping, trailer camp   ¶NEARBY ATTRACTIONS: Indian Burial Cave, Fantasy World Caverns, Bridal Cave, Lake of the Ozarks State Park, Bagnell Dam resort area.

MARK TWAIN CAVE

Mailing address: P.O. Box 822, Hannibal, MO 63401   *Phone:* (314) 221-1656
Directions: On State Highway 79, about 1 mile south of Hannibal.

Located in the bluffs along the Mississippi River, Mark Twain Cave is the actual cave that influenced the description by Samuel Clemens when he wrote his classic *Tom Sawyer*. Few caves have had such an historian, and although the tale was fictitious, the story is part of American culture. Tom and Becky, who were lost here, have thrilled generations of young people. It is fortunate that Twain had seen this cave, for it embodies all of the mystery and confusion that his story relates. The visitor today can as easily become lost in the maze-like intersecting corridors; he can readily imagine the plight of the children, without a guide and with no other light than candles and matches as they wandered from room to room.

Mark Twain Cave, Missouri.

The cave is dry and the floor nearly level—not much changed since the days when Twain visited it. An indirect lighting system has been installed, but during summer months at 7:30 and 8:00 P.M. there is a Candlelight Tour, when the electricity is turned off and visitors carry candles. There is little likelihood that visitors will see Injun Joe, but the flickering candlelight does make the story more believable.

¶OPEN: daily all year 8:00 A.M. to dusk   ¶GUIDED TOUR: 1 hour; the evening Candlelight Tour, 1¼ hours   ¶ON PREMISES: snack bar, gift shop, camping, trailer camp, picnicking   ¶NEARBY: motel, hotel, restaurant   ¶NEARBY ATTRACTIONS: Cameron Cave, Mark Twain Historic District, Hannibal, Mark Twain Birthplace State Park, riverboat trips.

MARVEL CAVE

Mailing address: Silver Dollar City, MO 65616   *Phone:* (417) 338-2611   *Directions:* On State Highway 79, about 12 miles west of Branson.

Marvel Cave, located about three miles from White River in a beautiful section of the Ozarks, has become the number one cave attraction in Missouri. The story of this accomplishment is the story of a remarkable family who acquired the lease to the cave in 1950 and with diligence, industry, and good taste built Silver Dollar City above the cave that now attracts over a million and a half visitors per season. The cave has enjoyed an attendance almost to its capacity for the past ten years. Today, the cavern is not featured in advertising for Silver Dollar City, as it has become only one of many attractions for the visitor to enjoy.

Originally called Devil's Den, the cave was visited by Osage Indians as early as 1500. Later, in the 1800s, it was named Marvel Cave when it was mined for the bat guano contained in the entrance room. After the fertilizer was mined out, Marvel was exhibited as a public show cave until 1950 by the Lynch family. Mary and Hugo Herschend of Chicago leased the cave at that time and with their two sons began the work which resulted in the present park.

Marvel Cave has one of the largest entrance rooms in the United States. The sinkhole drops directly into a huge chamber which, as shown below, is large enough to launch a hot air balloon. Visitors descend a wooden fire tower in the entrance chamber and follow a steeply sloping mountain trail from the base of the tower to the floor of the room. The trail then continues down into the lower level of the cave where a waterfall can be seen. An inclined railway has been installed at the end of the tour to eliminate the need to retrace steps on the return journey.

The theme of Ozark life in the 1880s above the cave has been continuously expanded, and now there is a resident group of craftsmen who carry on the art and trades of that period, providing visitors with a view of and participation in a living museum. Admission to the cave is part of a ticket that visitors buy which includes the many tastefully executed attractions of this theme park.

Marvel Cave, Missouri.

¶OPEN: April through October. Summer daily, 9:00 A.M. to 7:00 P.M. Spring and fall (closed Monday and Tuesday) 10:00 A.M. to 6:00 P.M.   ¶GUIDED TOUR: 1 hour   ¶ON PREMISES: All facilities for overnight are outside the park, but all facilities for a whole day's visit are available in the park.   ¶NEARBY ATTRACTIONS: Talking Rocks Cavern, Cosmic Cavern (Arkansas), Onyx Cave (Arkansas), Fantastic Caverns.

MERAMEC CAVERNS

Mailing address: Stanton, MO 63079   *Phone:* (314) 468-4156   *Directions:* Take Exit 230 off Interstate 44 at Stanton.

The bluffs of limestone along the Meramec River in central Missouri contain a number of caves. One of these, known as Salt Peter Cave, was mined at the time of the Civil War for saltpeter used in the making of gunpowder. Later it was the scene of country dances for the local folk as well as a favorite picnic spot. In 1933 a man named Lester Dill purchased the cave and changed the name to Meramec Caverns. He also changed the sleepy, bucolic spot into one of the best-known tourist attractions in Missouri. Under Dill's diligent care, the cave was publicized, promoted, and became prosperous. Today visitors walk into a spacious entrance and tour the cavern as seen through the eyes of this master showman. Visitors may not believe all the tall tales and legends about the cave, but it is fun, and the formations and underground river are of interest. There are large formations, large rooms, and a cave well worth seeing.

¶OPEN: daily all year. Summer 8:00 A.M. to dusk; winter 9:30 A.M. to dusk ¶GUIDED TOUR: 1 hour   ¶ON PREMISES: restaurant, snack bar, gift shop, camping, trailer camp, picnicking   ¶NEARBY: motel at La Jolla Natural Park ¶NEARBY ATTRACTIONS: Onondaga Cave, Missouri Caverns, Meramec State Park, Washington State Park.

Meramec Caverns, Missouri.

ONONDAGA CAVE

Mailing address: Leasburg, MO 65535   *Phone:* (314) 245-6600   *Directions:* 65 miles southwest of St. Louis on Interstate 44. Take Exit 215.

The Meramec River Valley was a favorite hunting ground for Daniel Boone, and if we can believe the legends, he discovered Onondaga Cave in 1798. The natural entrance at that time contained a small stream exiting from a narrow

Onondaga Caverns, Missouri.

tunnel. It was necessary to wade, crawl, and swim for several hundred feet to reach the main passages. In the early 1800s millers dammed Lost River coming from the entrance; this prevented further exploration of the cave. It was not until 1886 that further exploration was made and the cave opened to the public. In 1904, visitors from St. Louis came by rail and carriage to see the wonders of the cave. And wonders they still are. The huge rooms are profusely decorated and glistening, for the cave is still active and growing.

Onondaga Cave is recognized as one of the finest show caves in Missouri and also one with the most troubled history. For many years there was a land dispute regarding ownership. This was "settled" by each owner exhibiting his end of the cave, the two properties having been separated by a barbed wire fence. Advertised as two caves, Onondaga and Missouri Caverns were in fierce competition for many years. All this is now forgotten, as both are now under a single ownership. The latest threat to the cave was from the Army Corps of Engineers and a plan to build a dam that would flood the cave. This project, deferred at present, would bring about the loss of one of the most beautiful caves in central United States.

¶OPEN: all year 9:00 A.M. to dusk   ¶GUIDED TOUR: 1 hour and 20 minutes ¶ON PREMISES: restaurant, snack bar, gift shop, camping, trailer camp, picnicking ¶NEARBY: motel, hotel ¶NEARBY ATTRACTIONS: Meramec Caverns, Meramec State Park, Fisher's Cave, Meramec Springs, Davisville State Park, Six Flags Over Mid-America.

OZARK CAVERNS

Mailing address: Osage Beach, MO 65065   *Phone:* (314) 346-2500   *Directions:* State Road A off U.S. 54, about 4 miles south of Osage Beach or 4 miles east of Camdenton.

The Big Springs Valley near the Lake of the Ozarks has recently been made part of the State Park system, and Ozark Caverns will be included in this control beginning in the 1979 season. The cave was developed in 1952. Formerly named Coakley Cave, it was well known by local people. The addition of trails, lights, and handrails has made it possible for visitors to view it with little effort. The entrance room extends as a broad flat-ceilinged passage for several hundred feet back into the valley slope. A narrow meandering passage leads to the main trunk passage where an actively growing stalactite cluster provides a near-waterfall called the "Angel's Shower Bath." The entire trail covers almost 1400 feet of passageway and returns to the entrance through the same narrow sinuous route.

¶OPEN: April to October, 9:00 A.M. to dusk   ¶GUIDED TOUR: 45 minutes ¶ON PREMISES: gift shop, picnicking   ¶NEARBY: motels, restaurant, Osage Beach resort area   ¶NEARBY ATTRACTIONS: Bridal Cave, Indian Burial Cave,

Jacob's Cave, Lake of the Ozarks State Park, Bagnell Dam resort area, Fantasy World Caverns.

OZARK UNDERGROUND LABORATORY

Mailing address: Protem, MO 65733   *Phone:* (417) 785-4289   *Directions:* Directions for locating this cave can be obtained by writing to the mailing address.

One of the most unusual cave developments in the United States is the Ozark Underground Laboratory located in a remote area near the small town of Protem, Missouri. Not a tourist cave in the regular sense, it is available to groups for the study of the cave environment. Thomas Aley, Director of the Laboratory, has prepared a most unusual and instructive course that consists of both a surface and underground tour for groups. The surface tour takes about two hours, and the lecturer shows the relationship of the surface to the cavern development. The underground tour takes 2½ to 3 hours depending upon the group size. The cave is privately owned, and most visitors are associated with schools or colleges. The field trips are designed to complement and enhance high school and college courses in the physical, biological, and earth sciences. Serious researchers are encouraged to contact the laboratory for further information regarding the facilities and programs offered.

The cave trip is along a 2100-foot trail which has been installed to serve two functions: to make the tours relatively easy to negotiate, and—more importantly—to prevent disturbance of the floor and walls of the cave beyond, which are used for particular studies. There are no electric lights; flashlights and lanterns are used. Named Tumbling Creek Cave, it has five times more passage than is exhibited. One area is off limits, as it contains a colony of 150,000 gray bats (*Myotis grisenscens*), one of the few remaining retreats for these endangered creatures. There are approximately 70 animal species known in the cave, eight of these new or undescribed, including a new genus of cave snail. The cave is an educational experience, and it is closely protected to provide a fine opportunity for those who appreciate the cavern environment.

¶OPEN: by appointment only   ¶GUIDED TOUR: surface, 2 to 2½ hours; underground, 2½ to 3 hours. No one is permitted in the cave without supervision. ¶ON PREMISES: field house with electricity, running water, stove, and refrigerator. Groups supply tents, sleeping bags, and their own food. Camping is encouraged, as one or two nights would provide the best opportunity to complete the field trips and studies.   ¶NEARBY ATTRACTIONS: Bull Shoals Lake (5 miles), Table Rock Lake, Silver Dollar City, Talking Rocks Cavern, float trips.

OZARK WONDER CAVE

Mailing address: Route 2, Noel, MO 64854   *Phone:* (417) 475-3579   *Directions:* 4 miles north of Noel and ½ mile east of State Route 59 at Elk Springs.

Located in a long narrow hill on the edge of Elk River, Ozark Wonder Cave as it is now known was reputedly discovered by Confederate troops who were building a road across the ridge to Elk Spring. Later limestone quarrying

Ozark Wonder Cave, Missouri.

operations removed part of the cave, leaving an opening in the cliff face. As a result, some of the old formations which were previously inside the cave are now seen on the vertical exterior wall. The passages follow a joint pattern that parallels the hill in a nearly straight line.

The cavern was only a local curiosity until J. A. Truitt decided that the fine formations would be of interest to visitors, and in 1916 he constructed trails and stairways. "Dad" Truitt, as he was called, was something of a local legend for he was responsible for the development of five show caves in southwest Missouri. The interest that Truitt had in caves extended to the archaeological remains found there. He contacted the Museum of the American Indian in New York City, and for several seasons excavations in local caves, including Ozark Wonder Cave, turned up artifacts and information regarding the early inhabitants of the region.

¶OPEN: all year daily 8:00 A.M. to 6:00 P.M.   ¶GUIDED TOUR: 45 minutes ¶ON PREMISES: gift shop, cabins, picnicking, resort operated by cave owners ¶NEARBY: motel, hotel, camping, trailer camp, swimming, fishing and all resort facilities   ¶NEARBY ATTRACTIONS: Crystal Caverns, Bluff Dwellers Cave, Roaring River State Park, Onyx Cave (Arkansas).

REBEL CAVE

Mailing address: Rebel Cave Road, Patterson, MO 63956   *Phone:* (314) 224-3795   *Directions:* Junction of U.S. Highway 67 and State Highway 34.

Many cave owners in southeast Missouri claim that Jesse James and his band used their cave as a hideout. While Rebel Cave does not claim this, it might have been possible. There were many "rebels" living in this sparcely settled area of Missouri at the end of the war acting almost independently as guerrillas and harassing the flanks of the Union Forces. One such group might have occupied this cave for a time, as there is a mass grave marker in a local cemetery for seven men who were executed on May 28, 1865, nearly six weeks after General Lee's surrender. One of these men was a grandson of the owner of Rebel Cave.

The cave entrance is in a large sinkhole with nearly vertical walls, the cave extending into and beneath the hill on one side. The present tour is self-guided with a taped narration providing information at each stopping point. Just inside the entrance is a natural chimney to the surface that has been the natural ventilator for those who previously sought refuge in the cave. Excavations in the entrance show that the cave was used by Indians; fire pits and charcoal attest to many years of occupation. Several rooms, unknown to early visitors, have been opened by digging and are now exhibited.

¶OPEN: April through October, daily 9:00 A.M. to 5:00 P.M.   ¶SELF-GUIDED TOUR: Cave lighted with electric lights. Narration provided by push-button tape

Rebel Cave, Missouri. Photo by M. Jefferis.

recordings.   ¶On premises: gift shop, trailer camp, picnicking   ¶Nearby: restaurant, camping, motels, hotels   ¶Nearby attractions: Gad's Hill (site of Jesse James train robbery), Big Spring State Park, Cave Spring Onyx Caverns, Sam Baker State Park, Wappapello and Clearwater Lakes.

ROUND SPRING CAVE

Mailing address: Ozark National Scenic Riverways, Box 490, Van Buren, MO 63965   *Phone:* (314) 323-4236   *Directions:* 330 yards off State Highway 19.

The entrance to this cave is only a few feet above Spring Valley Creek and is part of the protected Ozark National Scenic Riverways. Opened to the public in 1930, it was displayed as a private enterprise until taken over by the National Park Service and is now shown by Park Rangers to groups of limited size. The narrow winding tunnel that leads into the cave extends about 400 feet to a point where it enters the main trunk passage in approximately the center of the corridor. The trail leads left and right through a cave averaging 25 feet wide and 20 feet high for a distance of approximately ½ mile in either direction. Several rooms are more than 100 feet wide—enlargements of the general winding tunnel. A stream traverses much of the main passage. The cave contains large formations and features areas of aragonite crystals, helictites and sponge work. There are no electric lights; the tour is lighted by lanterns provided by the rangers.

¶OPEN: Memorial Day to Labor Day. Daily 11:00 A.M. and 5:00 P.M (two tours only). Friday, Saturday, and Sunday an extra tour at 2:00 P.M.   ¶GUIDED TOUR: 2 hours. Lanterns provided. Reservations required. Limited to ten participants.   ¶ON PREMISES: camping, trailer camp, picnicking   ¶NEARBY: motel, restaurant (Eminence, 12 miles)   ¶NEARBY ATTRACTIONS: The Sinks, Montauk State Park, Ozark National Scenic Riverways, float trips, swimming, fishing.

TALKING ROCKS CAVERN

Mailing address: Silver Dollar City, MO 65616   *Phone:* (417) 272-3366   *Dirmctions:* On State Highway 13, 5 miles south of Reeds Spring.

Formerly known as Fairy Cave, this cavern was exhibited with lanterns and equipped with ladders in the 1920s for hardy and energetic visitors. Later, in 1929, masonry stairways were installed. Essentially one large joint-controlled chamber, the cave extends about 225 feet into a ridge, averages about 40 feet wide, and is over 100 feet deep. Entrance is made from the top of a chamber, and a stairway descends through many formations to the bottom of a canyon and then retraces to the entrance. Renamed Talking Rocks Cavern when the cave changed ownership, a light and sound experience has been added to the tour. The massive decorations and fine acoustics of the chambers provide a dramatic and exciting cave trip.

¶OPEN: May through October. Summer daily, 9:00 A.M. to 6:00 P.M. Spring and fall, closed Thursday and Friday   ¶GUIDED TOUR: 35 minutes   ¶ON PREMISES: snack bar, gift shop, picnicking   ¶NEARBY: restaurant, camping, motels, hotels, trailer camps   ¶NEARBY ATTRACTIONS: Marvel Cave, Silver Dollar City, Crystal Caverns, Table Rock Lake resort area.

THE SINKS

Mailing address: Gladden Star Route, Salem, MO 65560   *Phone:* (314) 858-3371   *Directions:* Off State Highway 19 about 10 miles north of Round Spring on County Routes EE and CC.

This private recreation area includes campsites, fishing ponds, and swimming holes along Sinking Creek. The natural feature that makes it part of this guidebook is a natural bridge over Sinking Creek. About 400 feet long, it spans the creek, and visitors may boat through the grotto. The water is clear, and the sunlight reflecting through the water lights the inner chambers, making the water seem blue. Flat-bottomed boats are available to make the short journey and reservations may be made at the Cabin Store located at the campgrounds.

Talking Rocks Cavern, Missouri.

The Sinks, Missouri.

¶OPEN: Memorial Day to Labor Day during daylight hours   ¶GUIDED TOUR: with flat-bottomed boats, about 15 minutes   ¶ON PREMISES: snack bar, gift shop, groceries, camping, trailer camp, swimming, and fishing   ¶NEARBY: motel, restaurant   ¶NEARBY ATTRACTIONS: Round Spring Cave, Round Spring State Park, Welch Spring, Ozark National Scenic Riverways, Alley Spring, Blue Spring.

MONTANA

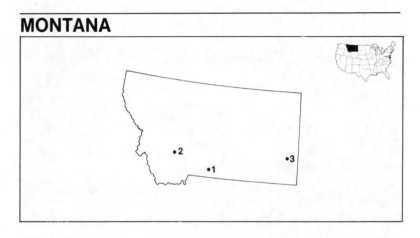

1 Big Ice Cave 2 Lewis and Clark Caverns State Park 3 Medicine Rocks State Park

THE GREAT WESTERN PLAINS extend through Montana into Canada. Surfaced with the rock waste of the Rocky Mountains and earlier volcanic debris, the eastern portion of the state is relatively level. The western boundary of the plains is dramatically outlined by the jagged skyline of the Rocky Mountains. The uplift which formed the Rockies also folded and faulted the limestones of the region. Since this rock is widely distributed in Montana, there are many potential areas for cave discovery. The caves now open to the public are varied in origin and features.

BIG ICE CAVE

Mailing address: Route 2, Box 110, Red Lodge, MT 59068   *Phone:* (406) 446-2103   *Directions:* From Bridger, 35 miles east off U.S. Highway 310 on unnumbered County Road and Forest roads in the Pryor Mountains unit of Custer National Forest.

This has to be one of the more remote show caves of the United States. Located in the beautiful Pryor Mountains, it can be found only with the aid of a guide.

Visitors who wish to see this unusual ice cave should make contact with the ranger station at the above address. The cave, located in a lovely alpine meadow, has a small entrance passage that at certain times of the year is completely choked with ice. The entrance room contains remnants of ice stalactites until Labor Day, but most of the ice is gone in late summer. A boardwalk has been built over the ice by the Forest Service and a spiral staircase has been installed in order to gain access to a lower room. A jacket is recommended as the cave temperature rarely exceeds 35° F. The Pryor Mountains have been the range of wild horses for more than a hundred years and a trip to the cave might be enhanced by a view of these splendid animals, although they are wary and probably cannot be approached.

¶OPEN: Saturday, Sunday, and holidays, July 4 to Labor Day, 10:00 A.M. to 4:00 P.M.   ¶ON PREMISES: picnicking   ¶NEARBY: nearest facilities 35 miles away   ¶NEARBY ATTRACTIONS: Bighorn Canyon National Recreation area, Bighorn Lake, Old Fort Smith, Chief Plenty Coups Memorial State Monument.

LEWIS AND CLARK CAVERNS STATE PARK

Mailing address: P.O. Box 1024, Three Forks, MT 59752   *Phone:* (406) 285-3694   *Directions:* Off U.S. Highway 10 about 10 miles south of Interstate 90. Take either Cardwell or Three Forks Exits.

The Jefferson River valley in southwestern Montana was the trail for the first expedition headed by Lewis and Clark when they made their journey to the Pacific in 1805. Although they passed only two miles from the cave, it was not until 1892 that deer hunters discovered the entrance. Known for many years as Morrison Cave, it was made a National Monument in 1908 by President Theodore Roosevelt, and called Lewis and Clark Caverns in honor of the two early explorers.

The tour starts from a Visitors' Center; a ¾-mile paved walk leads to the cave entrance. The underground tour is another ¾ of a mile, and since visitors exit through another tunnel, it is only one mile back to the Center. The cavern has a vertical, rather than a horizontal, profile; visitors enter the highest point of the cave and descend to a lower level, where a 500-foot horizontal man-made tunnel avoids the climb back out. This is the largest show cave in Montana and is profusely decorated with mature formations. The setting of the park and trails is spectacular. The parking lot, at 5300 feet, provides a fine view of the surrounding country.

¶OPEN: May 1 to September 30, 9:00 A.M. to 5:00 P.M.   ¶GUIDED TOUR: 90 minutes. Allow an extra half-hour for walking the trail to the cave.   ¶ON PREMISES: snack bar, gift shop, camping, trailer camp, picnicking, nature trail, campfire programs   ¶NEARBY: restaurant, hotel (25 miles)   ¶NEARBY attractions: Missouri Headwaters State Park, Madison Lake, Hot Springs, Boseman Hot Springs, deep copper mines.

MEDICINE ROCKS STATE PARK

Mailing address: Ekalaka, MT 59324   *Phone:* (none listed)   *Directions:* Off State Highway 7 about 20 miles south of Baker.

This 220-acre park contains strange weathered pinnacles of sandstone that are riddled with wind-blown caves and hollows. There are no caves long enough to require lights to explore, but the curious erosion that has formed these spires is of interest. The hollows in the pinnacles were once the meeting place of early Indian tribes. An exhibition of the history of the region can be found in Ekalaka Museum, 10 miles to the south.

¶OPEN: all year   ¶SELF-GUIDED TOUR: A flashlight might be advisable for that possible "dark cave."   ¶ON PREMISES: camping and picnicking   ¶NEARBY: all facilities in Baker or Ekalaka   ¶NEARBY ATTRACTIONS: O'Fallon Historical Society Museum, Ekalaka Museum.

NEBRASKA

THERE ARE NO CAVES open to the public in the state of Nebraska.

NEVADA

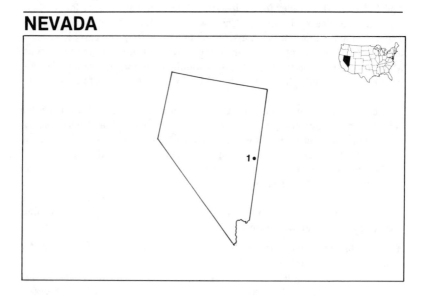

1 Lehman Caves National Monument

NEVADA IS LOCATED IN THE Great Basin area between the Rocky Mountains and
the Sierra Nevadas. The mountain ranges that run north-south in parallel ridges
through much of the state were formed by a series of faults that created block
mountains with short steep slopes on one side and a long gentler slope on the
other. Erosion has softened some of the sharp ridges but the outlines are still
evident. Some of the limestone exposed by this erosion is ancient, perhaps 550
million years old, but the caves are relatively recent in origin. The caves of the
region never exceed a mile in total length, a characteristic of highly fractured
limestone.

LEHMAN CAVES NATIONAL MONUMENT

Baker, NV 89311   *Phone:* (702) Lehman Caves Toll Station #1   *Directions:* 5
miles west of Baker at end of Nevada State Highway 74.

Lehman Caves National Monument, Nevada.

Located on the eastern flank of Wheeler Peak on the Nevada-Utah border, Lehman Cave is one of the highest show caves in the United States. The natural entrance, discovered by Absalom Lehman in 1885, is nearly 7000 feet above sea level. The trails and exit tunnel provide an easy underground tour in spite of the elevation. The cave is noted for its fine collection of palettes or shields — disk-shaped calcite formations — that are abundant throughout. These are relatively rare in the caves of this country and theories as to their origin are still disputed by geologists. While Lehman Cave is not large (the entire portion exhibited could fit within half of the Big Room in Carlsbad Caverns), the variety and abundance of formations make this one of the most colorful and decorative of the caves of the west.

¶OPEN: all year 8:00 A.M. to 5:00 P.M.   ¶GUIDED TOUR: 1 hour and 30 minutes ¶ON PREMISES: snack bar, gift shop, picnicking   ¶NEARBY: restaurant, camping, motel, trailer camp   ¶NEARBY ATTRACTIONS: Wheeler Peak scenic area, Copper Pit (one of the largest glory holes in the mining world), Ward Charcoal Ovens (approximately 50 miles).

NEW HAMPSHIRE

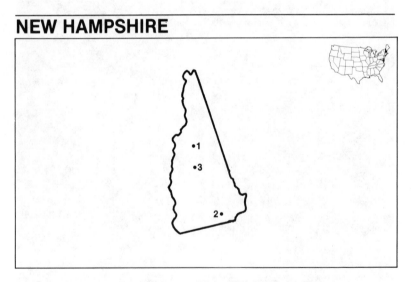

1 Lost River Reservation 2 Mystery Hill (artificial) 3 Polar Caves

NEW HAMPSHIRE IS SITUATED in the New England Highlands, with the Adirondack Mountains to the west and the Atlantic Ocean lapping the shores of the southeast corner. Battered at least four times by ice floes during the glacial period, the once-jagged mountains of the state have been sheared and softened in outline by ice, wind, and weather. There is no limestone in New Hampshire, but the public has access to many interesting natural crevices, shelters and boulder collapses which show the action of the glaciers.

LOST RIVER RESERVATION

Mailing address: Box 87, North Woodstock, NH 03262 *Phone:* (603) 745-8031
Directions: 6 miles west of North Woodstock, State Highway 112.

Lost River was so named because the brook draining the southern part of Kinsman Notch disappears below the surface in a narrow, steep-walled glacial gorge. This gorge is partially filled with immense blocks of granite under which the brook cascades along its subterranean course until eventually it emerges and joins the Pemigewasset River. The chaotic jumble of large boulders, deposited

Lost River Reservation, New Hampshire. Photo by Dick Hamilton.

by the action of moving glacial ice, provides the caves that are a feature in this area. The angularly shaped blocks have been eroded by the abrasive action of sand and gravel which were carried along by the glacial melt water. Potholes, cut by the same process, are found throughout the gorge: one of these is 28 feet in diameter and more than 60 feet deep.

The self-guided trail follows a gorge through the boulders, spans the gorge on wooden bridges, and in places ascends by ladders to the various levels.

¶OPEN: mid-May to mid-October, 9:00 A.M. to 6:00 P.M.   ¶SELF-GUIDED TOUR: Bring lights for the caves and travel at your own pace (takes approximately 45 minutes to an hour).   ¶ON PREMISES: snack bar, gift shop, picnicking   ¶NEAR-BY: restaurant, motel, camping, hotel, trailer camp   ¶NEARBY ATTRACTIONS: Old Man of the Mountain State Park, Mitersil ski area, Wildwood State Park, The Flume, Franconia Notch State Park, Indian Head, Polar Caves.

Mystery Hill, New Hampshire.

MYSTERY HILL (Artificial caves)

Mailing address: 29 Highland Avenue, Derry, NH 03038 *Phone:* (603) 893-8300 *Directions:* Take Exit 3 off Interstate 93, 5 miles east on State Highway 111.

These artificial caves are included in this guide because they were known locally as Pattee's Caves and have a curious and controversial origin. In the 1930s Jonathan Pattee excavated these "caves" for the purpose of finding out their archaeological background. By disturbing this archaeological site, he may have removed all possibility of ever discovering the true history of these unusual structures, which are a series of rock rooms, many with roofs of huge stones. Some are now exposed to the sky. Much has been written about the past history of these "caves," with archaeological claims ranging from theories of ancient Eskimo dwellings that could have been built when the Eskimos were forced south by the Ice Age, to a theory, presented in book form, that the site was inhabited and built by Irish Culdee monks fleeing from Viking raiders in 900 A.D. There is also speculation that there may be an association with the fogous (underground rooms) of Cornwall, as they resemble these structures.

¶OPEN: April to December 1, weather permitting. Spring and fall, 10:00 A.M. to 4:00 P.M.; summer, 9:00 A.M. to 5:30 P.M. ¶GUIDED TOUR: 2 hours ¶ON PREMISES: snack bar, gift shop, picnicking ¶NEARBY: restaurant, motel, hotel, camping ¶NEARBY ATTRACTIONS: Benson's Animal Farm, Canobie Lake park, Rockingham Race Track, Kingston State Park.

POLAR CAVES

Mailing address: R.F.D. 2, Box 180, Plymouth NH 03264 *Phone:* (603) 536-1888 *Directions:* Take Exit 26 on Interstate 93 at Plymouth, about 5 miles west on State Route 25.

About 15,000 years ago the glaciers retreated from the area now known as New Hampshire and deposited a "jack-straw" jumble of granite boulders at the base of the cliff face of Mount Haycock. The boulders were not dashed to the base of the cliff but were gently deposited by the melting ice, which also left crevices, cracks, and passages beneath the rocks. These interstices are now known as Polar Caves.

The tour of the area passes through six caves beneath these granite blocks. All have been given descriptive names such as Nature's Ice Box, King's Wine Cellar, Devil's Turnpike, and Smuggler's Cave. Because the base rock is granite, there are no formations such as one sees in most of the limestone caves of the country. There are, however, interesting mineral exposures of quartz, feldspar, mica, garnet, and rare deposits of beryl as well as limestone dissecting many of the huge granite boulders.

Polar Caves, New Hampshire. Photo by Mildred A. Beach.

¶OPEN: mid-May to mid-October 9:00 A.M. to 5:00 P.M.   ¶SELF-GUIDED TOUR: Caves are lighted; follow the trails. About 1 hour   ¶ON PREMISES: cafeteria, snack bar, gift shop, picnicking, waterfowl exhibit, maple sugar museum, mineral display   ¶NEARBY: camping, motel, trailer camp, restaurant, golfing and horseback riding   ¶NEARBY ATTRACTIONS: Lost River, Loon Mountain, Clark's Trained Bears, The Flume, Castle in the Clouds, White Mountain resort areas, Cardigan State Park, Wellington State Park, Squam Lake.

NEW JERSEY

THE ONLY CAVES IN New Jersey are found in the northwest part of the state in the Piedmont Plateau. South Jersey is in the Coastal Plain and has no caves. There are no show caves in the state.

NEW MEXICO

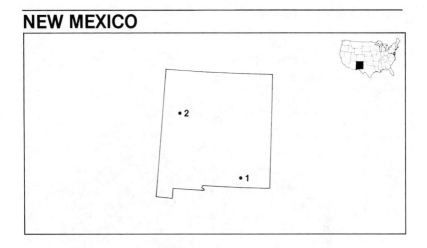

1 Carlsbad Caverns National Park 2 The Desert Ice Box

THE STATE OF NEW MEXICO is divided by the southern end of the Rocky Mountains. To the west lies the Colorado Plateau and to the east lies the southern part of the Great Western Plains. The plateau area shows evidence of recent lava flows and volcanic activity. The southeast plains are broken by uplifted limestones that outcrop in the Guadalupe Mountains.

Two hundred and fifty million years ago a limestone reef formed in a shallow sea. When the uplifts of the Rocky Mountains occurred, the reef was raised, forming the Guadalupe Mountains. The largest known room in an American cave is found in this reef—in Carlsbad Caverns. Some of the lava flows also contain caves.

In past times the Indians of New Mexico used the shallow sandstone caves in the plateau area to build their remarkable cliff dwellings. These caves have been omitted from this guide at the request of the Park Service. Only serious researchers are permitted access to these most fragile remains, as they could not survive heavy tourist traffic.

CARLSBAD CAVERNS NATIONAL PARK

3225 National Parks Highway, Carlsbad, NM 88220   *Phone:* (505) 885-8884
Directions: About 20 miles south of Carlsbad on U.S. 62 & 180 at White City take State Highway 7 west to the cave.

This is the most celebrated and famous cavern in the United States. Located 32 miles from the town of Carlsbad, it has no overnight accommodations, no railroad, no airport and is miles from the main east-west highway routes that cross the country. Nevertheless, more than 20 million people have visited this remote spot since the cave was opened to the public by a visionary young cowboy named James Larkin White in the early 1920s. Interest in the cave led

Carlsbad Caverns National Park, New Mexico.

what was then a private enterprise to become a National Park in the 1930s. Extensive development since then has made it the most important underground visitor attraction in the United States. Its natural beauty includes a spectacular display of giant formations, the largest single room in an American cave, and a fortunate natural layout which makes a dramatic presentation for visitors.

Every visitor who takes the complete tour walks into a large and impressive cave entrance. He descends on an easy sloping trail that switches back on itself as he goes deeper and deeper into the cave. Even after a descent of more than 500 feet, the light from the entrance is still visible. The rooms are huge and newly lighted with an excellent system. The Big Room is awe-inspiring. The easy trail that winds around the room requires more than an hour to traverse and off in an alcove of the cave is a lunchroom large enough to feed several hundred people. There are also restroom facilities here. In the early days of the exhibition of the cave this was the end of the tour and the visitors retraced their steps to the surface, but today there are two high-speed elevators which bring parties up to the Visitors' Center in a matter of seconds. No one leaves this cave without the realization of its immense size and depth beneath the earth. The elevators also permit visitation to the cave by the disabled, as the trails below will accommodate wheelchairs.

One unusual feature is the evening bat flight which occurs at approximately 7:00 P.M. during the summer. Countless millions of Mexican free-tailed bats have used a portion of this great cave for thousands of years. At sunset the bats begin to emerge, swirling and milling about as they spiral out and fly off in a seemingly endless stream. The flight varies with weather conditions, but at times several million bats exit the cave, returning singly throughout the night and before dawn. These beneficial animals consume tons of insects. They are harmless to man, and are protected against disturbance during their migratory stay at the cave.

There are more than 50 natural caves within the park; however only the main cavern and New Cave are open to visitors. New Cave is undeveloped and requires some strenuous hiking and walking. It is located 23 miles from the Visitors' Center. Explorers must provide transportation to the entrance, and should bring a flashlight, sturdy shoes, and drinking water. Reservations are required a day in advance, and the trip is limited to 25 persons. It is an adventure well worth the effort, as New Cave contains huge formations and spectacular rooms. It is a taste of wild cave exploration under competent supervision.

¶OPEN: all year 8:00 A.M. to 4:30 P.M. except June through September when the hours are 7:00 A.M. to 8:00 P.M. New Cave: daily trips in summer; after September on weekends only. Time of trips will vary and information is available by calling (505) 785-2233 ¶SELF-GUIDED TOUR: Carlsbad Caverns: Allow three hours. Each visitor is given a radio receiver that provides taped narration as he proceeds at his own pace through the cave. (Narrations are available in adults' and children's version, in English and in Spanish.) No one is allowed to leave the trails, and rangers are stationed throughout the cave to give interpretive information to visitors and to protect the fragile formations. ¶GUIDED TOUR: New Cave: Allow four hours from Visitors' Center. Cave is one mile from the New Cave parking lot (and 500 feet up a switch-back trail). The hike to the cave takes approximately 45 minutes; since tour departures are scheduled from the New Cave entrance, allow sufficient time to hike up the mountain to the grouping point. Cave tour duration is about 1½ hours. Reservations must be made one day in advance. Phone (505) 785-2233 ¶ON PREMISES: restaurant, snack bar, gift shop, picnicking ¶NEARBY: motels, camping and trailer camps. There are no facilities at New Cave. ¶NEARBY ATTRACTIONS: Guadalupe Mountains National Park (Texas), Living Desert State Park, Sitting Bull Falls, Potash mines.

THE DESERT ICE BOX

Mailing address: Box 12000, Grants, NM 87020 *Phone:* (505) 783-5194 *Directions:* Off State Highway 53 about 20 miles southwest of Grants.

The Desert Ice Box is a cave in a lava flow that centuries ago turned to stone. A wall of ice of unknown depth forms a semicircle closing off the back of the cave. At the top of the stairs leading down to the ice cave, the temperature may reach 100° F. Seventy-five feet below, visitors step into a 30° F. refrigerator. Natural, perpetual ice has been insulated from the outside heat by ropy layers of lava. Near the cave, Indian trails and stone tools have been found dating back approximately 1000 years.

¶OPEN: all year, weather permitting   ¶GUIDE SERVICE OR SELF-GUIDED: natural lighting   ¶ON PREMISES: snack bar, gift shop, camping, cabins, trailer camp, motor tours, ski-run in season, picnicking   ¶NEARBY: scattered facilities ¶NEARBY ATTRACTIONS: Great Malpais Lava Flow, Bandera Crater, Indian Ruins, El Morro National Monument, Ramah Ruins (Zuni Indian village), Acoma Indian Reservation, Laguna Pueblo and Mission, Chaco Canyon National Monument.

NEW YORK

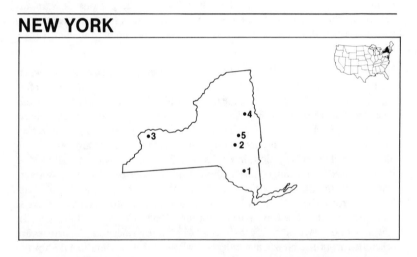

1 Howe Caverns 2 Ice Caves Mountain 3 Lockport Cave Raceway Tour (Artificial cave) 4 Natural Stone Bridge and Caves 5 Secret Cavern

THE STATE OF NEW YORK has a varied topography, from the sandy glacial moraine of Long Island on the eastern Atlantic Coast to the natural boundary of the Great Lakes on the north and west. The middle of the state is divided by the Appalachian Plateau, and the Adirondack Mountains skirt the eastern border. Most of the caves are found in the central region where limestones 375 million years old lie in gently folded beds. Many caves were crushed or buried by the repeated ice flows during glacial times. Only narrow fissure-type passageways were able to support the enormous weight of the ice. Historically the caves were used for shelter, but the evidence of prolonged occupancy by Indians is limited to shallow entrances.

HOWE CAVERNS

Mailing address: Howes Cave, NY 12092   *Phone:* (518) 296-8990   *Directions:* Off State Highway 7 between Central Bridge and Cobleskill.

The largest and most celebrated cave in the northeast, Howe Caverns is located in the heart of the cave country of New York State. Although the original entrance was known to Indians, it was not until Lester Howe donned his exploring clothes, pushed back into the deeper recesses of the cave, and publicized his discovery that the cave was visited by the public. Opened first as a local curiosity, it soon achieved national fame in pre-Civil War days. A hotel was built, and dances and weddings were held underground. At times Lester Howe would personally entertain his friends with violin concerts in the resonant chambers. But economic problems caused the business to fail, and the cave was closed for many years. The quarrying of limestone at one end of the cave removed part of the entrance area, and it seemed that Howe's cave was to be forever forgotten. However, in 1929 a group of local businessmen put down an elevator shaft into the area that was originally the rear of the cave tour and opened the portion still visited today. This includes a stream passage with

Howe Caverns, New York.

many formations, an underground lake with boat ride, and an interesting sinuous winding passage. The main passageway of Howe Caverns is remarkable in New York State for its size, and this in combination with a tasteful development and imposing setting on the surface make it a major cavern attraction.

¶OPEN: all year 9:00 A.M. to 6:00 P.M.   ¶GUIDED TOUR: 1¼ hours, includes ¼-mile underground boat ride   ¶ON PREMISES: restaurant, motel (open six months), snack bar, hay ride, gift shop, picnicking   ¶NEARBY: all facilities   ¶NEARBY ATTRACTIONS: Old Stone Fort, Secret Caverns, Toe Path Mountain State Park, Thompsons Lake State Park, Lower Adirondacks, Baseball Hall of Fame (Cooperstown), John Boyd Thatcher State Park, many lakes and reservoirs.

ICE CAVES MOUNTAIN

Mailing address: Cragsmoor, NY 12420   *Phone:* (914) 647-7989   *Directions:* About 4 miles south of Ellenville on State Highway 52, turn 1 mile north.

The Shawangunk Mountains near Ellenville, New York, are the remnants of once higher mountains that have been eroded by wind, ice, and water. The bedrock atop the mountain has been fractured by earth movements, leaving a series of crevices and cracks bridged over by glacial debris to make caves and shelters near the cliff face. One of these crevices has been opened as a visitor attraction providing a self-guided tour through the breakdown. In the spring and summer, laurel and wild flowers bloom along the walking trails. Winter snows accumulate in the crevices to form ice that remains frozen until late in the summer.

¶OPEN: April through November weather permitting. Daily 9:00 a.m. to dusk   ¶SELF-GUIDED TOUR: Three-mile loop drive from entrance gate to cave. Parking areas at Sam's Point Lookout and Ice Caves. About ½-mile walk. Caves are lighted. Allow 1 to 1½ hours   ¶ON PREMISES: snack bar, gift shop, picnicking   ¶NEARBY: motel, camping, trailer camps, restaurant   ¶NEARBY ATTRACTIONS: New Paltz Stone Houses, Monticello Raceway, Catskill Mountains resort area, Old Rhinebeck Aerodrome, Hyde Park, Vanderbilt Mansion, Senate House (Kingston).

LOCKPORT CAVE RACEWAY TOUR (artificial cave)

Mailing address: 32 Cherry Street, Lockport, NY 14094   *Phone:* (716) 433-9027   *Directions:* Near downtown Lockport on State Highway 78, north at Clinton Street in Upson Park.

Lockport "Cave" is an artificial tunnel blasted out of rock to provide a by-pass for the water used in the Erie Canal. Engineered by Birdsill Holly in1858, it was an important part of the canal system that linked the east coast with the middle west. The era of the canal was cut short by railroad competition, and the Lockport tunnel was adapted as a raceway for power. In 1909, industrial use of the water was abandoned and the tunnel was sealed beneath the ruins of an old pulp mill and its presence almost forgotten. In 1973 the young enthusiast Thomas Callahan learned of the "cave" and set out to locate it. He excavated and explored the tunnel, finding it safe and accessible to traverse. With the aid of local historians and town officials he opened the cave as the Lockport Cave Raceway Tour. Today the Raceway features historical ruins plus geologic cave formations that are beginning to form in the man-made tunnel. A walk along the Erie Canal and the Lockport Locks provides a glimpse of the historic relics of the Industrial Revolution as guides furnish explanation, including the folklore and history of the area.

¶OPEN: May through October, 10:30 A.M. to 5:00 P.M.   ¶GUIDED TOUR: about 1 hour and 20 minutes. Includes underground boat ride and tour of Erie Canal and Lockport Locks   ¶ON PREMISES: picnicking only   ¶NEARBY: all facilities   ¶NEARBY ATTRACTIONS: Niagara Falls, Lake Ontario, Lake Erie, Fort Niagara, Tonawanda Indian Reservation, Tuscarora Indian Reservation.

NATURAL STONE BRIDGE AND CAVES

Mailing address: Pottersville, NY 12860   *Phone:* (518) 494-2283   *Directions:* Take Exit 26 off Interstate 87 on Stone Bridge Road about 2 miles west.

In the heart of the Adirondacks is the largest state park in the United States, with nearly six million acres of private and public land only a day's drive from New York and Boston. Natural Stone Bridge and Caves lie within this Adirondack State Park but are privately owned and operated. The Natural Bridge has been a landmark near Pottersville since Jacob VanBenthuysen built a mill on the stream in 1790. The remains of the mill can still be seen, as subsequent generations of the VanBenthuysen family have preserved both the mill site and the natural aspects of the area. There is a series of caves, grottoes, potholes, tunnels, waterfalls, rapids, cliffs, and ledges in this rustic setting. There are at least seven caves, including Barrel, Echo, Garnet, Geyser, Kelly Slide, Noisy, and Lost Pool. All of them have pools, running streams, or waterfalls. The caves are easily traversed by adults or children. A self-guiding tour map is given each visitor to enable him to see the entire area.

¶OPEN: May through October, 8:00 A.M. to dark   ¶SELF-GUIDED TOUR: about 1 hour; caves are lighted   ¶ON PREMISES: snack bar, gift shop, trout fishing, rock and mineral shop, Mermaid Swim show   ¶NEARBY: restaurants, camping, motels, cabins, trailer camp, resort area   ¶NEARBY ATTRACTIONS: Lake Champlain, Schroon Lake, Lake George, Paradox Lake, Crown Point,

Natural Stone Bridge and Caves, New York.

Ticonderoga, Frontier Town, Fort William Henry, and many other Adirondack Mountain attractions.

SECRET CAVERNS

Mailing address: P.O. Box 88, Cobleskill, NY 12043   *Phone:* (518) 234-3431
Directions: About 35 miles west of Albany, between U.S. 20 and State Highway 7.

Secret Caverns shows many features of a typical New York state cave. It has sinuous passageways with high domes, an active stream, and evidence of glacial action in the remains of gravel washed through the cave. Its natural entrance permits the visitor to see the beds of the overlying limestone and the casts of fossils embedded in the stone. Developed in 1929 by Roger H. Mallery, the portion of the cave exhibited is only a small part of the underground drainage system of the area. A waterfall in the cave demonstrates the water action that scoured out the walls and carved the erratic shapes left in the limestone.

¶OPEN: April through October. During months of May, June and September, 9:00 A.M. to 7:00 P.M.; July and August open 8:00 A.M. to 9:00 P.M.; balance of season 9:00 A.M. to 5:00 P.M.   ¶GUIDED TOUR: 45 minutes   ¶ON PREMISES: gift shop, camping, picnicking   ¶NEARBY: all facilities   ¶NEARBY ATTRACTIONS: Old Stone Fort, Howe Caverns, Lower Adirondacks, Baseball Hall of Fame (Cooperstown), John Boyd Thacher State Park, Schuyler Mansion.

NORTH CAROLINA

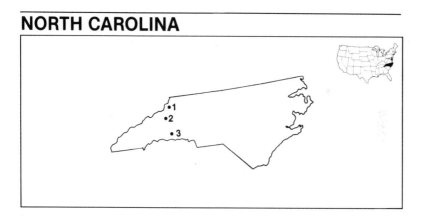

1 Chimney Rock Park 2 Grandfather Mountain 3 Linville Caverns

THERE ARE THREE DISTINCT regions of North Carolina: the flat sandy coastal plains to the east, the Piedmont Plateau in the middle of the state, and the Appalachian Mountains to the west. The first two areas have no caves, but small pockets of limestone in the Appalachians have a few caves. The cracks and faults in the ancient rock also provided some shelters and overhangs called "caves" by the early settlers. Once the hunting grounds of Cherokee Indians, the mountains are now a popular resort area for travelers along the Eastern Seaboard.

CHIMNEY ROCK PARK

Mailing address: Chimney Rock, NC 28720 *Phone:* (704) 625-9204

Directions: Off U.S. 74-64 and State Highway 9, about 22 miles southeast of Asheville.

Chimney Rock Park is a scenic area of massive rocks, waterfalls, nature trails, and panoramic views. A tunnel blasted 196 feet through solid granite leads to an elevator that takes visitors to the top of Chimney Rock. A minor part of the area is Moonshiner's Cave. A stairway leads down to where there was once a still.

¶OPEN: daily March through November; summer from 8:30 A.M. to 5:30 P.M., winter from 8:30 A.M. to 4:30 P.M.   ¶SELF-GUIDED TOUR: no light required ¶ON PREMISES: snack bar, gift shop, picnicking, nature trails, 400-foot waterfall ¶NEARBY: camping, motels, hotels, cabins, trailer camp   ¶NEARBY ATTRACTIONS: Museum of North Carolina Minerals, Mount Mitchell State Park (highest point east of the Mississippi), Cowpens National Battlefield Site (South Carolina), Biltmore House and Gardens (Asheville), Linville Caverns (45 miles), Grandfather Mountain.

GRANDFATHER MOUNTAIN

Mailing address: Linville, NC 28646   *Phone:* (704) 733-2800   *Directions:* Off U.S. 221 about 3 miles north of Linville.

Located atop one of the highest mountains in the Blue Ridge are several interesting crevices and shelters. Black Rock Cliffs Cave is a fissure extending several hundred feet into the cliff. Indian House Cave is a shelter about 70 feet long with a history of Indian occupation. Both of these caves are on the property of Grandfather Mountain Recreation Preserve and can be reached by marked foot trails from the Visitors' Center. A private scenic roadway permits visitors to drive to the top of the mountain where there are facilities for camping and picnicking. A popular area for group gatherings, there are accommodations for business and church meetings. The caves are only a small part of the attractions of the park.

¶OPEN: from April 1 to November 15; spring and fall, 8:00 A.M. to 6:30 P.M., summer from 8:00 A.M. to 8:30 P.M.   ¶SELF-GUIDED TOUR: flashlights recommended   ¶ON PREMISES: snack bar, gift shop, picnicking, hiking, bear habitat, hang gliding, Highland Games (second week in July), Gospel Singing on the Mountain   ¶NEARBY: camping, motels, hotels, cabins, trailer camp   ¶NEARBY ATTRACTIONS: Museum of North Carolina Minerals, Mount Mitchell, Linville Caverns, Land of Oz, Chimney Rock Park.

LINVILLE CAVERNS

Mailing address: P.O. Box 567, Marion, NC 28752   *Phone:* (704) 756-4171 *Directions:* On U.S. 221, about 18 miles north of Marion.

While hunting in the Linville Valley more than a hundred years ago, H. Colton and Dave Franklin found the entrance to what is now known as Linville Caverns, at the foot of Humpback Mountain. Intrigued by the fish in the stream issuing from the entrance, they explored the cave with torches. They described their discovery to friends saying that it looked like the arches of some grand old cathedral. Later the cave was used as a hiding place by deserters from both armies in the War Between the States. In it an old man made and mended shoes for the soldiers. Today a highway goes directly past the entrance, and a parking area has been carved out of the hill to accommodate visitors. The trail follows an underground stream which has been stocked with trout. Although the floor is level throughout, interesting convolutions of the walls and ceilings provide a sinuous twisting corridor between the rooms. The cave is well decorated with formations.

¶OPEN: daily all year 8:30 A.M. to 6:30 P.M. in summer; 8:30 A.M. to 5:00 P.M. in winter   ¶GUIDED TOUR: 35 minutes   ¶ON PREMISES: gift shop, picnicking ¶NEARBY: restaurants, snack bar, camping, motels, hotels, cabins and trailer camps   ¶NEARBY ATTRACTIONS: Grandfather Mountain, Blowing Rock, Chimney Rock Park, Mount Mitchell State Park (highest point east of the Mississippi), Museum of North Carolina Minerals, Blue Ridge Parkway.

NORTH DAKOTA
THERE ARE NO show caves in North Dakota.

OHIO

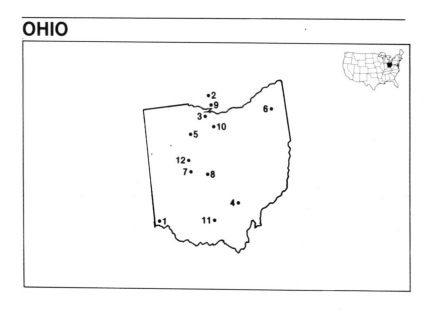

1 Cincinnati Museum of Natural History (artificial) 2 Crystal Cave (Put-In-Bay) 3 Crystal Caves Park (Sandusky) 4 Hocking Hills State Park 5 Indian Trail Caverns 6 Nelson's Ledges 7 Ohio Caverns 8 Olentangy Indian Caverns 9 Perry's Cave 10 Seneca Caverns 11 Seven Caves 12 Zane Caverns

THE MISSISSIPPI VALLEY PLAINS extend to the Great Lakes and include the area of Ohio. This nearly level region is underlain by limestones 300 to 350 million years old. Numerous caves are found where this limestone has been randomly exposed throughout the state. Occupied by Indians for more than 4000 years, the area was thoroughly explored. Nearly every cave and shelter shows evidence of Indian visitation.

CINCINNATI MUSEUM OF NATURAL HISTORY (artificial cave)

Mailing address: 1720 Gilbert Avenue, Cincinnati, OH 45202   *Phone:* (513) 621-3889   *Directions:* Downtown Cincinnati near the Art Museum.

This remarkable artificial cave was designed by Ralph Ewers, a dedicated speleologist, while a technician at the Cincinnati Museum. There are 400 feet of meandering passages including displays of delicate formations, a 40-foot waterfall, and hundreds of tiny synthetic cave creatures placed in life-like positions. This educational display has proved to be the most popular exhibit at the museum with more than 20,000 adults and children "exploring" it each year.

¶OPEN: all year Monday through Saturday 9:00 A.M. to 5:00 P.M.; Sunday 1:00 P.M. to 5:00 P.M.   ¶SELF-GUIDED TOUR: 20 minutes   ¶ON PREMISES: gift shop ¶NEARBY: all facilities in downtown Cincinnati.

CRYSTAL CAVE

Mailing address: Put-In-Bay, South Bass Island, OH 43456   *Phone:* (419) 285-2811   *Directions:* South Bass Island is reached by ferry from Port Clinton on State Highway 2, or Catawba Point via State Highway 53. Island Airlines serves the Island from Port Clinton.

Crystal Cave, on the property of the Heineman Winery and Vineyards, has a solid mass of strontium crystals that cover much of the interior of the cave. These crystals, the largest of their kind in the world, have ten faces. The angle of each face on one crystal is exactly the same as that on the similar face of every other crystal. This deposit of strontium sulfate is the only one of any size in the United States. The cave was discovered in 1897 while a well was being dug.

¶OPEN: Memorial Day to Labor Day, 11:00 A.M. to 5:00 P.M.   ¶GUIDED TOUR: about 15 minutes   ¶ON PREMISES: wine shop, gift shop, snack bar, picnicking ¶NEARBY: scattered facilities   ¶NEARBY ATTRACTIONS: Perry's Cave, Seneca

Crystal Cave, Ohio.

Caverns, Crystal Caves Park, Perry Memorial National Monument, Blue Hole, Kellys Island State Park, Catawba Island State Park, Birthplace of Thomas A. Edison.

CRYSTAL CAVES PARK

Rural Route 2, Box 51, Sandusky, OH 44870   *Phone:* (419) 684-7177   *Directions:* Off U.S. 6 about 7 miles west of Sandusky.

This small cave was once used as a water supply for the Crystal Rock Beer Company of Sandusky. The clear water from the limestone was reputedly piped all the way to the city. Today the entrance of the cave is enclosed in a modern building that covers the original sinkhole with a stairway leading from the lobby down to the spring. A taped narrative provides a description of the cave and the formations. The tour ends at the bottom of the sinkhole within the same building. There is another cave on the property that was open to the public, but which was fenced off because of excessive vandalism. Crystal Caves Park functions primarily as a campground and picnic spot for visitors to the Cedar Point and Lake Erie Island resort areas.

¶OPEN: Memorial Day to Labor Day, 9:00 A.M. to 5:00 P.M.   ¶SELF-GUIDED tour: about 30 minutes. Taped narration at station points throughout the cave   ¶ON PREMISES: snack bar, gift shop, camping, motel, trailer camp, picnicking, meeting hall for 200 persons   ¶NEARBY: restaurant, hotel and all facilities in Sandusky (7 miles)   ¶NEARBY ATTRACTIONS: Blue Hole, Seneca Caverns, Crystal Cave (Put-In-Bay), Thomas A. Edison Birthplace, Perry's Cave.

HOCKING HILLS STATE PARK

Mailing address: Route 2, Logan, OH 43138   *Phone:* (614) 385-6841   *Directions:* On State Route 664 and 374, about 12 miles southeast of Logan. Park spreads over 5-by-10-mile area. See Park Office maps.

This beautiful park has many scenic attractions, including waterfalls and three caves open to the public. Ash Cave, while small, is known for its acoustical properties which magnify even the slightest sounds. The cave received its name from ashes of fires believed to have been kindled by Ohio's early settlers. Old Man's Cave is more of a gorge than a cave, and includes interesting vegetation and cascades. Rock House is situated in a 150-foot high cliff. Large pillars of weathered sandstone in a variety of colors grace the cave entrance halfway up the cliff. The cave has seven passageways, and is reputed to have been used as a shelter for Indians, a cache for horse thieves, and a hideaway for fugitives.

¶OPEN: all year, daylight to dark   ¶SELF-GUIDED TOUR: 1 to 3 hours   ¶ON PREMISES: restaurant (March 1 to November 30), snack bar, gift shop, camping, cabins, trailer camp, fishing, swimming, horseback riding, hiking   ¶NEARBY: scattered facilities   ¶NEARBY ATTRACTIONS: Tar Hollow State Park, Conkels Hollow State Park, Cedar Falls State Park, Mound City Group National Monument, Lake Logan State Park, Lake Hope State Park, Indian Mounds, Leo Petroglyphs.

INDIAN TRAIL CAVERNS

Mailing address: Box 127, Vanlue, OH 45890   *Phone:* (419) 387-7015   *Directions:* About four miles northwest of Carey on State Highway 568.

Indian Trail Caverns, located on the border of the old Big Spring Indian Reservation, was undoubtedly known to the early Wyandotte Indians. In 1927 the cave was partially excavated of glacial debris and opened to the public. The venture failed in 1937, but in 1973 the cave was reopened with trail improvements and electric lighting. The cave is only a few feet below the surface, a sinuous horizontal corridor, with several collapsed "skylights" where entrance to the main passage is possible. One such opening has a natural stone ladder that might have been used by the Indians to gain access to the cave.

¶OPEN: weekends only Memorial Day to September 30, 1:00 P.M. to 6:00 P.M. Daily by appointment. Phone in the evening (419) 387-7773   ¶GUIDED TOUR: 45 minutes to an hour   ¶ON PREMISES: picnicking   ¶NEARBY: all facilities in Findlay (10 miles)   ¶NEARBY ATTRACTIONS: St. Mary's Shrine, Wyandotte Indian Mission Church, Hull Old Trail Monument, Van Buren State Park, Seneca Caverns, Hancock County Museum.

NELSON'S LEDGES

Mailing address: R.F.D. #2, Box 292, Garrettsville, OH 44231   *Phone:* (216) 548-8348   *Directions:* 3 miles east of Parkman on State Route 282.

Nelson's Ledges lies partly within Nelson-Kennedy Ledges State Park. The park's huge rocks, caves, and constantly flowing streams are the result of an ancient glacier which, as it moved, carved deep gorges and created interesting formations. A moraine of boulders more than a mile long has been deposited, forming many boulder caves. The longest are Gold Hunter's Cave and Devil's Den. This is a wild jumbled place with interesting areas to explore. No special equipment is required except sturdy shoes and flashlights.

¶OPEN: March 1 to November 7, 9:00 A.M. to dark   ¶SELF-GUIDED TOUR: flashlights advisable in two main caves   ¶ON PREMISES: restaurant, snack bar, gift shop, picnicking   ¶NEARBY: all facilities on U.S. 422   ¶NEARBY ATTRACTIONS: First Morman Temple in U.S., McKinley Memorial, Garfield Birthplace, Punderson State Park, many lakes and reservoirs.

OHIO CAVERNS

Mailing address: R.F.D. #1, West Liberty, OH 43357   *Phone:* (513) 465-4017 *Directions:* 3 miles southeast of West Liberty on State Route 245.

Ohio Caverns was discovered in 1897 by Robert Noffsinger, a young farm worker, when he decided to find where the water drained from a low spot in a field in the Mingo Valley of west central Ohio. He dug only a few feet before he found a crevice which led to the caverns. Opened to the public that same year, the cave enjoyed moderate success until 1925 when new discoveries permitted the opening of the section presently shown to the public. The original area discovered by Noffsinger is not now exhibited.

Ohio Caverns has remarkable pure white formations. Since the ceiling of the cave is generally low, a trench has been dug in many places for visitors to walk. This provides a comfortable viewing height across the flat bedding planes within the cave. The whiteness of the formations is contrasted by the dark limestone and the reddish clay of the floor. Most of the formations are still active and growing, with glistening reflections from the wet surfaces. The one-way, mile-long tour exits through an artificial tunnel.

Ohio Caverns, Ohio.

¶OPEN: all year 9:00 A.M. to 5:00 P.M.   ¶GUIDED TOUR: 1 hour   ¶ON PREMISES: snack bar, gift shop, picnicking   ¶NEARBY: camping, restaurant, motel, trailer camp   ¶NEARBY ATTRACTIONS: Zane Caverns, Olentangy Caverns, Mac-O-Chee Castle, Indian Lake State Park, Kiser Lake State Park.

OLENTANGY INDIAN CAVERNS

Mailing address: Delaware, OH 43015   *Phone:* (614) 548-7917   *Directions:* About 8 miles north of Columbus off U.S. 23.

Central Ohio was the hunting area for the Wyandotte Indians; and Olentangy Caverns was used as a refuge and ceremonial site before the coming of the white man. Evidence within the cave shows that the Indian Council Room was used by Chief Leatherlips and his braves as a workshop for making arrowheads and other stone implements as late as 1810. Some of these artifacts are on display at the Cave House on the surface. The earliest record of visitation by white men was in 1821 when an early pioneer, J. M. Adams, camped near the entrance of the cave while enroute to the west. During the night one of his oxen strayed into the sinkhole and was killed. Mr. Adams found the dead beast at the bottom of

Olentangy Caverns, Ohio.

the hole and explored partway into the cave. He left his name and date on the wall where it can still be seen today.

Entrance is by means of a stairway that descends this sinkhole to the second level where a maze of passages lead off to a series of rooms. There are four levels in the cave. The lowest has an underground river that drains into the Olentangy River a half-mile away. This lowest level has been only partially explored.

¶OPEN: all year. April through October, daily, 9:30 A.M. to 5:30 P.M.; November through March, weekends only   ¶SELF-GUIDED TOUR: 25 to 30 minutes with taped narration at station points throughout the cave   ¶ON PREMISES: snack bar, gift shop, camping, trailer camp, picnicking   ¶NEARBY: all facilities   ¶NEARBY ATTRACTIONS: Zane Caverns, Ohio Caverns, Columbus Zoo, Delaware Dam and Reservoir, Delaware Lake State Park, Perkins Observatory, Leatherlips Monument.

PERRY'S CAVE

Mailing address: Put-In-Bay, South Bass Island, OH 43456   *Phone:* (419)

285-4081   *Directions:* South Bass Island is reached by ferry from Port Clinton on State Highway 2, or Catawba Point via State Highway 53. Island Airlines serves the Island from Port Clinton.

Perry's Cave is reputed to have been discovered by Commodore Oliver Hazard Perry in 1813. It is claimed that Perry used the cave to store arms before the Battle of Lake Erie and that British prisoners were kept there after the battle. The cave is 208 feet long and 165 feet wide.

¶OPEN: Memorial Day to Labor Day, 11:00 A.M. to 5:00 P.M.   ¶GUIDED TOUR: 15 to 20 minutes   ¶NOTE: Island can be reached by ferry. For schedule and rates, write: (at Port Clinton) Parker Boat Lines, Inc., Put-In-Bay, OH 43456 or phone (419) 732-22800; (at Catawba Point) Miller Boat Line, Inc., Put-In-Bay, OH 43456 or phone (419) 285-2421. By air via the Island Airlines, Box 172, Port Clinton, OH 43456 or phone (419) 734-3149   ¶ON PREMISES: gift shop, picnicking. Also on island: restaurant, snack bar, gift shops, state park, camping, hotel, motel, cabins, trailer camp, swimming, fishing   ¶NEARBY ATTRACTIONS: Crystal Cave, Seneca Caverns, Perry Memorial National Monument, Blue Hole, Kellys Island State Park, Catawba Island State Park, East Harbor State Park, Cedar Point, Birthplace of Thomas A. Edison.

SENECA CAVERNS

Mailing address: Bellevue, OH 44811   *Phone:* (419) 483-6711   *Directions:* 4 miles south of Bellevue on State Highways 18 or 269.

This unusual cavern in north central Ohio is popularly called "The Earthquake Crack" because it appears to have been formed by the shearing of the earth by massive forces into a nearly vertical crevice. The limestone has been clearly separated, and one wall would match the other if it were possible to rejoin them. The tour enters from the surface and descends to the water table nearly 150 feet below. A stone Indian rug needle found in the cave suggests that it was known some 2000 years ago. Legends among the tribes refer to the cave as sacred ground. Seneca Caverns was named for Seneca John, chief of the tribe that occupied a 15,000-acre reservation near the cave in the early days of white settlement. The word "Seneca" means "Place of Stone." Indians believed themselves to be the first people to appear on earth, and thought that their ancestors climbed up from vast subterranean caverns by means of a giant grapevine which had pierced the earth. A pool at the bottom of the cave, called Old Mist'ry River, is said to be connected to the artesian spring at Blue Hole 14 miles away. The only life found in this pool is a half-inch long shrimp that spends its entire life cycle in the darkness of the cave.

¶OPEN: daily 9:00 A.M. to 7:00 P.M. May 31 to Labor Day and weekends in May, September and October, 9:00 A.M. to 5:00 P.M.   ¶GUIDED TOUR: 1 hour

¶ON PREMISES: gift shop, camping, picnicking ¶NEARBY: restaurant, snack bar, motels, hotels, cabins, trailer camp ¶NEARBY ATTRACTIONS: Crystal Cave, Perry Memorial National Monument, Blue Hole, Kellys Island State Park, Crystal Caves Park, Birthplace of Thomas Edison, Catawba Island State Park.

SEVEN CAVES

Mailing address: Bainbridge, OH 45612   *Phone:* (513) 365-1283   *Directions:* Off U.S. 50 between Rainsboro and Bainbridge.

Located in the Paint Valley region of south central Ohio, Seven Caves is a private park that includes cliffs, canyons, and cascades in addition to the caves. Paved trails lead to the individual caves where push-button lighting permits the visitor to view them at his own pace. Each is different and although they are small, as remnants of a once larger cave system, they show the action of the water that formed them and the Rocky Fork Creek canyon that dissects them. Shelters are placed along the trail at advantageous viewing points. The park can be seen in an hour, but several hours can be spent picnicking and exploring in this pleasant area.

¶OPEN: March 15 to November 15 during daylight ¶SELF-GUIDED TOUR: about 1 hour; no lights necessary ¶ON PREMISES: restaurant, snack bar, gift shop, picnicking ¶NEARBY: camping, motels, cabins, trailer camp ¶NEARBY ATTRACTIONS: Seip Mound, Serpent Mound, Rocky Fork State Park, Pike Lake State Park, Alum Creek State Park, Ross Historical Museum.

ZANE CAVERNS

Mailing address: R.F.D. #2, Bellefontaine, OH 43311   *Phone:* (513) 592-6174 *Directions:* Off U.S. 33 on State Highway 540 about 7 miles east of Bellefontaine.

Zane Caverns is located near the highest point in the state of Ohio at the headwaters of Mad River. Discovered by John Dunlap in 1892, the cave has been open to the public since 1926 when two artificial entrances were dug to permit a one-way trip through the cave. There are two levels on the tour and in addition to fine dripstone formations there is a display of cave pearls.

¶OPEN: daily April through October 10:00 A.M. to 5:00 P.M. ¶GUIDED TOUR: 30 to 45 minutes ¶ON PREMISES: animal game park, snack bar, gift shop, camping, trailer camp, picnicking, playground ¶NEARBY: restaurant, motels, hotels, cabins ¶NEARBY ATTRACTIONS: Ohio Caverns, Olentangy Indian Caverns, Indian Lake State Park, Kiser Lake State Park, Mac-O-Chee Castle, Indian forts and blockhouses.

OKLAHOMA

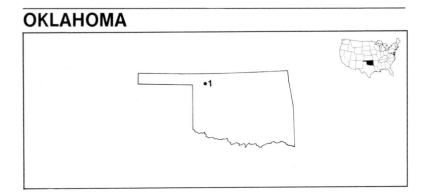

1 Alabaster Caverns State Park

THE GREAT PLAINS cover all of the state of Oklahoma except for the northeast corner which includes a small portion of the Ozark Mountains. Although there are several long limestone caves found in the state, none are open to the public. However, in interesting cave in gypsum is exhibited by the Oklahoma Park commission.

ALABASTER CAVERNS STATE PARK

Mailing address: Freedom, OK 73842   *Phone:* (405) 621-3381   *Directions:* 6 miles north of Freedom on State Highway 50.

At the edge of the Cimaroon River Valley in Cedar Canyon lies one of the most unusual show caves in the United States, Alabaster Caverns, formed in gypsum by the action of water. The entrance is in a rock cliff overlooking the canyon. The first chamber is more than 200 feet in diameter and 75 feet high. A trail winds through the massive rooms and emerges from a collapsed sinkhole on the edge of the canyon about a quarter of a mile away.

Gypsum is known as alabaster or calcium sulfate. In pure form it is colorless, but impurities cause it to assume different colors from pink to deep rust and white to dark gray. The pure form is also crystalline, providing masses of clear selenite up to two feet thick, that form on walls and ceilings of the cave. The individual crystals on the walls of Alabaster Caverns form a mosaic of interlocking design with some large crystals measuring 24 inches long by 12 inches wide.

The first recorded exploration was in 1900, but it was not until 1928 that the land was bought by Charles Grass. He encouraged visitors to bring lanterns and explore the underground rooms. In 1953 Grass transferred the ownership of the cave to the state of Oklahoma as a park.

¶OPEN: all year daily from 8:00 A.M. to 5:00 P.M.   ¶GUIDED TOUR: 45 minutes to 1 hour   ¶ON PREMISES: camping, trailer camp, picnicking, swimming pool, hiking   ¶NEARBY: most facilities within 10 miles   ¶NEARBY ATTRACTIONS: Boiling

Springs State Park, Little Sahara Recreation Area, Pioneer Museum, Freedom Museum.

OREGON

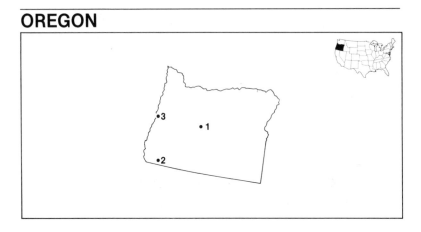

1 Lava River Caves State Park 2 Oregon Caves National Monument 3 Sea Lion Caves

THE STATE OF OREGON is divided into three topographic regions: the narrow coast range to the west, the Columbia Plateau to the east, and the Cascade Mountains in the middle. The extensive lava beds on the eastern slope of the Cascades have many lava tube caverns. Limestones are scattered about the state, but only in the southwest and northeast are they cavernous. The coastal region has sea caves and natural arches formed by the action of the waves.

LAVA RIVER CAVES STATE PARK

Mailing address: 211 N. Revere Bend, OR 97701   *Phone:* (503) 382-5668
Directions: On U.S. 97 about 12 miles south of Bend.

The U.S. Forest Service maintains a Visitors' Center on U.S. Route 97 near Bend, Oregon which is the starting point for an interesting overview of the lava beds that dominate the landscape of the area. A large raised topographic map at the Center provides direction for the visitor, and a computerized sound and film presentation presents the features of the park. The State of Oregon also administers some of these lands. Lava River Cave in particular is only a mile from the Center and provides an exciting experience in a huge, mile-long tunnel through the lava. This excellent example of a lava tube has no guide service, but a ranger on duty will rent gasoline lanterns to visitors for a nominal fee. It is strongly recommended that an alternate source of light be taken in case of failure of the lanterns. The cave temperature is a brisk 41° F., providing an exhilarating walk. The black walls seem to absorb the lantern light, and the shadows it casts create fanciful changing images.

There are other caves in the park, three of them open to visitors: Skeleton Cave, Charcoal Cave, and Lavacicle Cave. The last requires written permission to enter, and guides from the Forest Service will lead tours by reservation only. It is recommended that anyone wishing to follow the foot trails and visit the caves check with the Lava Lands Visitors' Center for trail directions.

¶OPEN: May through September, 9:00 A.M. to 4:30 P.M.   ¶SELF-GUIDED TOUR: about 1 hour; lanterns available but bring extra lights   ¶ON PREMISES: there are no facilities at the caves, but the Visitors' Center has a small gift shop and bookstore   ¶NEARBY: all facilities are in Bend (12 miles)   ¶NEARBY ATTRAC-TIONS: Newberry Crater, Lava Cast Forest, Arnold Ice Cave, Lapine State Park, Pilot Butte.

OREGON CAVES NATIONAL MONUMENT

Mailing address: Oregon Caves, OR 97523   *Phone:* Oregon Caves Toll Station No. 1   *Directions:* About 50 miles south of Grants Pass, take U.S. 199 to Cave Junction, then State Highway 46 to Oregon Caves.

Oregon Caves (it should be singular—there is only one cave) was discovered in the fall of 1874 by Elijah Davidson while hunting deer on the slopes of the Siskiyou Mountains. His dog chased a bear into the entrance of a cave and Elijah followed, lighting matches to guide his way. He did not know that this chance discovery would engrave his name in the history books of the region. In 1909, local interests sent a petition to Washington, and President Taft proclaimed the area a National Monument. Today the cave is operated by a concessionaire, the Oregon Caves Company, under permit from the Department of the Interior. The cave consists of a series of narrow twisting passages that connect with intersecting rooms in an ascending direction. The tour traverses the cave, rising 200 feet in elevation and exiting from an artificial opening on the side of the canyon. The cave is profusely decorated and although not large has colorful and interesting formations. The area around the entrance has been improved with a rustic lodge and restaurant but the land is so steep and precipitous that there is no parking area for visitors. It is necessary to park several thousand feet away where the Park Service has cleared a level parking area. It is a beautiful setting with hiking trails maintained by the Park Service in the surrounding Siskiyou Forest.

¶OPEN: all year 8:00 A.M. to 5:00 P.M.   ¶GUIDED TOUR: 1½ hours. Tours usually consist of 12–16 visitors.   ¶ON PREMISES: restaurant, snack bar, gift shop, hotel, cabins, picnicking   ¶NEARBY: camping, trailer camp   ¶NEARBY ATTRACTIONS: Illinois River State Park, Siskiyou National Forest.

Oregon Caves National Monument, Oregon.

SEA LION CAVES

Mailing address: U.S. 101, Florence, OR 97439   *Phone:* (503) 547-3415   *Directions:* On U.S. 101 about 12 miles north of Florence.

The dramatic Oregon coastline with its spectacular waves and rocky promontories has an equally unusual feature in Sea Lion Caves. Located 12 miles north of Florence, it cannot be missed by the traveler taking the shore route. The highway bisects the area between the cave parking lot and the visitors' building; Pedestrian Crossing signs surprise the driver who has been traveling the Interstate Highways.

The visitor is treated to a most unusual display, for this is the only show cave containing sea lions available to American travelers. An enormous sea cave has been carved out by the relentless crashing of the surf on high volcanic cliffs. It is open on one side and leads to a 1500-foot-long tunnel. The opening to the sea permits daylight to illuminate what is the largest chamber of its kind on the North American continent. A colony of Stellar sea lions, some weighing as much as 2000 pounds, use this cave as a rookery during the mating and pup rearing period of the year. As many as 600 animals can be seen in the cave or on the rocks outside all during the year. The facilities for the visitor include graded walkways and sheltered areas with telescopes for viewing the sea lions during the spring and early summer. A high-speed elevator takes 50 seconds to descend 200 feet to another viewing platform where the animals can be seen within the cave (usually in winter and inclement weather). This is a privately owned attraction and stands as an excellent example of how concerned, motivated and sensitive people can exhibit a natural phenomenon in a tasteful and protective way.

¶OPEN: all year, June through September, 7:00 A.M. to dusk; October through May, 9:00 A.M. to dusk   ¶SELF-GUIDED TOUR: about 30 minutes. A guide is available at the cave viewing area to answer questions.   ¶ON PREMISES: gift shop, elevator, live sea lions   ¶NEARBY: all facilities on Route 101   ¶NEARBY ATTRACTIONS: rugged coastline, yachting, fishing. There are 15 state parks within 30 miles of the cave, most on the coast; Winchester Bay, Tenmile Lakes, Tahkenitch Lake, Siltcoos Lake, and others.

PENNSYLVANIA

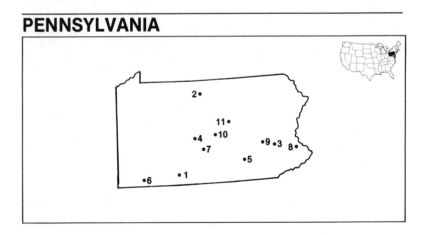

1 Coral Caverns 2 Coudersport Ice Mine (artificial cave) 3 Crystal Cave
4 Indian Caverns 5 Indian Echo Caverns 6 Laurel Caverns 7 Lincoln Caverns
8 Lost River Caverns 9 Onyx Cave 10 Penn's Cave 11 Woodward Cave

THE APPALACHIAN MOUNTAINS and the Appalachian Plateau extend diagonally
through the state of Pennsylvania. The mountains have been uplifted, folded,
and faulted several times during geologic history creating a complicated structure and topography. Although the caves of this area are not extensive, they
have features and beauty equal to any region. Many were known to the Indians
and many show evidence of their occupancy.

CORAL CAVERNS

Mailing address: Mann's Choice, PA 15550   *Phone:* (814) 623-6882   *Directions:* About 7 miles west of Bedford on State Highway 31 at Mann's Choice.

Coral Caverns, formerly called Wonderland Caverns, is located a short 500 feet
east of the main street of Mann's Choice. It was originally discovered by quarrying operations and is one of two caves broken into by the blasting. In 1932,
Philip Hughes installed stairways and lights in the larger of the two caves and
opened it to the public.

The cave has a fine display of fossils characteristic of the limestone of the
region. Stromataporoids, brachiopods, and horn corals predominate, all of
them casts of the original sea creatures that were imprisoned in the stone when it
was a sea bottom. The limestone has been tilted to a nearly vertical position so
that the wall of fossils is clearly displayed. The cave entrance is located near the
top of a hill. During the summer season a trackless train called the "Dinosaur
Express" takes visitors up the hill from a pavilion to the cave entrance.

¶OPEN: daily May 30 to Labor Day 10:00 A.M. to 6:00 P.M.   ¶GUIDED TOUR: 35
minutes   ¶ON PREMISES: gift shop, picnicking, trackless train to entrance
¶NEARBY: restaurant, motel, cabins, trailer camp   ¶NEARBY ATTRACTIONS:
Shawnee State Park, Magnesia Spring, Blue Knob State Park.

COUDERSPORT ICE MINE (artificial cave)

Mailing address: Coudersport, PA 16915   *Phone:* (none listed)   *Directions:*
Off U.S. 6 about 4 miles east of Coudersport.

Coudersport Ice Mine is located on Ice Mountain near the town of Sweden
Valley. Although it is artificial and not literally a cave at all, it is included here
because of interest it might have for visitors. Even the name "Ice Mine" is im-
proper because it is not a mine—ice is not removed from the pit. But ice is permitted to form in various shapes each season. The ice was discovered during dig-
gings for silver ore, and when the prospectors jokingly remarked that they had
found an ice mine, the name was born.

The Ice Mine is relatively free of ice in the winter. With the coming of spring — as early as March or as late as May — ice begins to form in the pit rapidly for 10 to 12 weeks. It remains throughout the summer until September or October when it begins to melt away. By November it is gone. This tiny pit is of interest as it is a true *glacière*, the most impressive one known in the eastern United States, and it follows the natural laws that have produced the great ice caves of Europe. Be warned however that a visit to the site merely provides a view of a rectangular hole in the ground with ice formations nearly filling it. The interest comes from the knowledge of the unusual conditions that have produced the ice.

¶OPEN: daily June, July and August 8:00 A.M. to 8:00 P.M. ¶GUIDED TOUR: 10 minutes ¶ON PREMISES: snack bar, gift shop ¶NEARBY: scattered facilities ¶NEARBY ATTRACTIONS: Sizerville State Park, Ole Bull State Park, Denton Hill State Park, Lyman Run State Park.

CRYSTAL CAVE

Mailing address: R.F.D. #3, Kutztown, PA 19530 *Phone:* (215) 683-6765 *Directions:* Between Allentown and Reading near Route 222, Kutztown.

In the fertile and prosperous farm country of Berks County, Crystal Cave was discovered in 1871 by William Merkel while quarrying limestone for the liming of his fields. Although not very long, it has large rooms. The geologic history of the

Crystal Cave, Pennsylvania.

cave is clearly shown by the nearly vertical limestone beds which have been tilted from the horizontal by great pressures from below. The cavern was formed by water flowing down this slope and attacked the weaker parts of the stone. The cave has been a tourist attraction for more than a hundred years. Its accessibility to New York City and Philadelphia makes it Pennsylvania's most popular cave. It is easily traversed along sloping concrete walks guarded by steel rails.

¶OPEN: from Washington's Birthday (in February) until the first Monday in December. February, March and April, daily 9:00 A.M. to 5:00 P.M.; May, 9:00 A.M. TO 5:00 P.M.; Saturdays, Sundays and holidays to 6:00 P.M.; Memorial Day through Labor Day, 9:00 A.M. to 6:00 P.M.; weekends to 7:00 P.M.; September to October 31st, 9:00 A.M. to 5:00 P.M., weekends to 6:00 P.M.; November, Friday, Saturday and Sunday only, 9:00 A.M. to 5:00 P.M.   ¶GUIDED TOUR: 35 minutes ¶ON PREMISES: gift shop, snack bar, farm animals, miniature golf, Indian tee-pees, nature trails ˙¶NEARBY: restaurants, motels, camping, trailer park, hotels ¶NEARBY ATTRACTIONS: Roadside America, Onyx Cave, W.K. & S. Steam Rail-road, Dutch Folk Festival, Hawk Mountain Sanctuary.

INDIAN CAVERNS

Mailing address: Spruce Creek, PA 16683   *Phone:* (814) 632-7578   *Directions:* Off State Highway 45 about 11 miles east of Tyrone between Waterstreet and State College.

Spruce Creek Valley in central Pennsylvania was a favorite hunting ground for the Iroquois and Algonquin Indian tribes. A pleasant stream in the valley had many campsites. One of them was just inside the entrance of a large cave only a few feet from the main trail. With the coming of the white man the trail became first a road, then a highway, and the history of the Indian occupation was forgotten. In 1928 Harold Wertz purchased the farm that included the cave and proceeded to explore the extensive cavern that led from the well-known entrance. Excavations in the first room disclosed artifacts that included more than 400 pieces of pottery, arrowheads, and clay pipes, as well as animal bones. The cave was opened to the public in 1929, and portions not visited by the Indians, including an underground stream, were added to the tour. Today the artifacts are on display. The areas of early visitation can be seen in the Indian Council Room and Writing Room where signatures of early explorers are still legible.

¶OPEN: all year daily 9:00 A.M. to dusk ¶GUIDED TOUR: 1 hour ¶ON PREMISES: gift shop, picnicking, trout fishing, hiking ¶NEARBY: restaurant, camping, motels, cabins, trailer camp ¶NEARBY ATTRACTIONS: Lincoln Caverns, Greenwood Furnace State Park, Whipple Dam State Park, Raystown Dam, East Broadtop Railroad, Horseshoe Curve, Pennsylvania Military Museum.

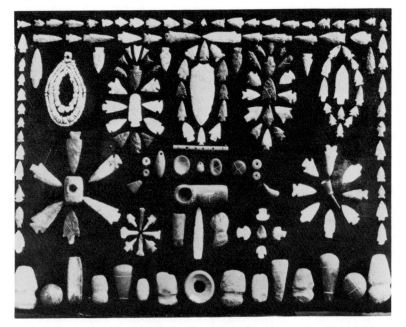

Indian Caverns, Pennsylvania.

Indian Echo Caverns, Pennsylvania.

INDIAN ECHO CAVERNS

Mailing address: P.O. Box 206, Hummelstown, PA 17036   *Phone:* (717) 566-8131   *Directions:* Off U.S. 322 about 10 miles east of Harrisburg.

The cave on the bank of the Swatara Creek was a local landmark, visited by Indians and later by white men who called it Echo Cave because of the acoustics in the entrance room. It was reputed to be sacred to the Susquehannock Indians who lived in the area and used it as a meeting place. In 1786 the cave became the home of a hermit named Amos Wilson. His only sister had been betrayed and abandoned by her lover. She was accused of killing her child, born out of wedlock, and was sentenced to be hanged. The Governor, petitioned by Amos Wilson, granted a pardon, but Wilson was delayed on his return from Philadelphia by swollen streams. When he reached the place of execution with the pardon it was too late; his sister was dead. The shock made him shun society and he lived as a recluse in the cave for nearly 20 years. His table and fireplace, shown above, are part of the tour, and the chimney and walls are exactly as he saw them 200 years ago. In 1929 the cave was opened to the public and explorations disclosed the undisturbed North Canyon where some of the finest formations are to be found. This area was made accessible by an artificial tunnel.

¶OPEN: March through November. Summer 9:00 A.M. to 7:00 P.M.; spring and fall 10:00 A.M. to 4:00 P.M.; March and November weekends only   ¶GUIDED TOUR: 45 minutes   ¶ON PREMISES: snack bar, gift shop, picnicking   ¶NEARBY: camping, motel, hotel, trailer camp, restaurant   ¶NEARBY ATTRACTIONS: Harrisburg, Gettysburg, Amish country, Hershey, Fort Hunter Museum, Carlisle Barracks, Molly Pitcher's Grave, Gifford Pinchot State Park, Lewis State Park, Pennsylvania Farm Museum, Pine Grove Furnace State Park.

LAUREL CAVERNS

Mailing address: P.O. Box 1095, Uniontown, PA 15401   *Phone:* (412) 329-5968   *Directions:* Off U.S. 40 east of Uniontown about 5 miles south of Mount Summit Inn.

Laurel Caverns is located atop Chestnut Ridge at the head of Cave Hollow in Fayette County. The entrance is now covered with a modern Visitors' Center where a masonry stairway leads directly into the cave. Known as Dulany's Cave after the original landowner who used to watch the horses of early visitors, it has been a favorite spot for adventurous local people since it was first described in 1816. It was given the present name of Laurel Caverns by Norman Cale, who opened the cave to the public in 1964.

Most of the more than seven miles of mapped passageway of this large cave is

Laurel Caverns, Pennsylvania. Photo by Charles Rosendale.

accessible for those who want the thrill of exploring by flashlights and carbide lamp. Part of the cave is a maze pattern, with passages forming a grid laid out like city streets. The lower part of the cave consists mainly of a broad, high, meandering passageway. The photo shows the self-guided Exploratory Tour to the undeveloped areas. The cave formed in limestone with a large amount of quartz or silica. The silica is not soluble in water and has been left in the passages to make a fill not unlike beach sand. The regular guided tour visits the upper maze part of the cave, which has a uniform tilt of about 15° from the horizontal. This slope adds interest even if it is initially confusing. There are few formations, but the opportunity to see a large cave in what is nearly a wild state makes this a worthwhile visit.

¶OPEN: daily all year 9:00 A.M. to 5:00 P.M. extended to 7:00 P.M. in summer ¶GUIDED TOUR: 45 minutes ¶EXPLORATORY TOUR: Between April 1 and November 15 there are daily self-guided tours to the undeveloped portions of the cave. These are spelunking tours and old clothes and a good flashlight with extra batteries will be needed. Carbide lights and a hard hat are recommended. A map of the cave will be furnished, and it will be the explorers' only guide with the exception of some key arrows in the cave. This tour takes about 2½ hours ¶ON PREMISES: snack bar, gift shop, picnicking, trackless train ¶NEARBY: camping, trailer camp, motel, hotel, restaurant ¶NEARBY ATTRACTIONS: Frank Lloyd Wright's Fallingwater, Fort Necessity National Battlefield, Albert Galatin Home, Braddock's Grave, Laurel Hill State Park, Kooser State Park.

LINCOLN CAVERNS

Mailing address: 5231 Simpson Ferry Road, Mechanicsburg, PA 17055 *Phone:* (814) 643-0268   *Directions:* On U.S. 22, 3 miles west of Huntingdon.

In 1930 when U.S. Highway 22 was being built, blasting along Warrior Ridge about three miles from Huntingdon opened up the entrance to Lincoln Caverns. Only 40 feet from the highway, it has attracted thousands of visitors. Continued exploration added the Mystery Room in 1937. In 1941, Myron Dunlavy, Jr. dynamited into a new section now called the Ancient Tomb. All of these discoveries are seen on the present tour, which covers nearly a half mile of trail. In spite of the violent blasting there is no damage within the cave, nor has it ever suffered from vandalism. The cave scenes are as pristine as when they were first exposed to the lights of the explorer.

¶OPEN: April through October 9:00 A.M. to 6:00 P.M.   ¶GUIDED TOUR: 1 hour ¶ON PREMISES: snack bar, camping, picnicking, gift shop ¶NEARBY: restaurant, camping, motels, hotels, cabins, trailer camp ¶NEARBY ATTRACTIONS: Swinart Automobile Museum, Raystown Lake, East Broadtop Railroad, Indian Caverns, Trough Creek State Park, Canoe Creek State Park.

LOST RIVER CAVERNS

Mailing address: Hellertown, Pennsylvania 18055   *Phone:* (215) 838-8767 *Directions:* About 3 miles south of Bethlehem on State Highway 412, at Hellertown turn ½ mile east.

In 1883, men quarrying limestone near Hellertown discovered a cave that became a local curiosity and picnic spot for nearly 50 years. Around the turn of the century a small dance platform was built inside the cave to take advantage of the cool air on hot summer nights. In 1930 the property was bought by E. C. Gilman, who excavated the entrance and made improvements. A stream unknown to early explorers was found that flowed for a short distance through the cave and then disappeared, giving rise to the name Lost River Cave. The cave is typical of those in the region with irregular winding passages and gentle grades. The highest ceiling is about 70 feet and the widest place about 60 feet. A solarium has been built over the entrance providing a glass-roofed enclosure where tropical plants are growing. A small museum and rock shop are operated as a continuing family enterprise.

¶OPEN: all year daily (except Christmas and New Year's Day) 9:00 A.M. to 8:00 P.M. ¶GUIDED TOUR: 30 to 40 minutes ¶ON PREMISES: snack bar, gift shop, camping (trailer only, no hookups), picnicking, wedding chapel in cave ¶NEARBY: all facilities ¶NEARBY ATTRACTIONS: Crystal Cave, Onyx Cave, Valley Forge, Hopewell Village National Historic Site, Ralph Stover State Park, Washington Crossing, Delaware Water Gap, Canal Museum, Liberty Bell Shrine.

ONYX CAVE

Mailing address: R.D. 2, Box 308, Hamburg, PA 19526 *Phone:* (215) 562-4335 *Directions:* Off State Highway 662 about 3 miles south of U.S. 22 in Hamburg.

Onyx Cave was discovered by quarrying operations — as were many others in the limestone regions of Berks County — in 1872, but it was not until 1923 that Irvin Dietrick opened it to the public. The cave consists of a nearly level passage that provides a comfortable walking tour through winding and sinuous passageways. The limestone beds are nearly flat, and the direction of the cave corridors is controlled by the vertical joint planes. A trail extends for almost 1000 feet through the cave.

¶OPEN: April 1 to October 31, 9:00 A.M. to 6:00 P.M. ¶GUIDED TOUR: 35 to 40 minutes ¶ON PREMISES: snack bar, gift shop, picnicking ¶NEARBY: restaurant, camping, motel, hotels, cabins, trailer camp ¶NEARBY ATTRACTIONS: Crystal Cave, Blue Rocks, Roadside America, Lost River Caverns, Valley Forge, Hopewell Village National Historical Site, French Creek State Park, Hawk Mountain Sanctuary.

PENN'S CAVE

Mailing address: Rural Delivery, Centre Hall, PA 16828 *Phone:* (814)

Onyx Cave, Pennsylvania.

364-1664   *Directions:* About 5 miles east of Centre Hall on State Highway 192.

Penn's Cave is the only show cave in Pennsylvania that is toured entirely by boat. Located in the Brush Valley of Centre County, it is the source of Penn's Creek, which rises from a spring in the entrance. Well known by the Indians, the caverns gave rise to many tales of romantic happenings. Land grants in

Penn's Cave, Pennsylvania.

1773 mention the cave as a landmark, but it was not until the middle 1800s that the underground stream was traversed by a raft. This soon became a popular local sport. In 1885 a hotel was built near the entrance to encourage visitors. The business was not successful, and in 1908 the family of the present owners acquired the land. Special craft powered by quiet outboard motors transport visitors more than half a mile through water-filled corridors, onto a small lake, and then back to the original entrance. The trip is an unusual experience that can be enjoyed without any walking or climbing. Penn's Cave Airpark, an asphalt airstrip on the property, permits visitors to come by private plane. For others there are plane rides available daily over the scenic Pennsylvania countryside.

¶OPEN: all year from 9:00 A.M. to 5:00 P.M. except from Memorial Day to Labor Day, when it is open until 8:00 P.M. ¶GUIDED TOUR: 1 hour (boat trip) ¶ON PREMISES: restaurant, snack bar, gift shop, camping, airport, picnicking ¶NEARBY: motel, hotel, cabins, trailer camp ¶NEARBY ATTRACTIONS: Woodward Cave, Poe Valley State Park, Bald Eagle State Park, Reeds Gap State Park, Boal Mansion, Lincoln Caverns, Indian Caverns.

WOODWARD CAVE

Mailing address: Route 45, Woodward, PA 16882   *Phone:* (814) 349-5185
Directions: About 1½ miles off State Highway 45 at Woodward, 25 miles west
of Lewisburg.

Woodward Cave, along Pine Creek in eastern Centre County, is named for the
nearby village. The large entrance shown below was well known to Seneca
Indians who, legends say, buried their dead in the cave. When it was first
explored, Pine Creek flowed through the cave in high flood times, although it
disappeared below ground before reaching the entrance in dry seasons. The
cave was filled with mud and driftwood brought in by each flood, and some
passages were completely choked with clay. In 1925 a group of local business
people built a diversion channel that carried the flood waters down the valley
by another course. They cleaned out the debris and installed walkways to
permit the easy tour now enjoyed in the cave. Since then there has been no
further flooding, and thousands of visitors have taken the half-mile circular
tour through the cave. The temperature within the cave is a constant 40° F.
Colorful Indian blankets are provided for those who do not come equipped
with a sweater.

¶OPEN: all year from 9:00 A.M. to 5:00 except from Memorial Day to Labor
Day when it is open 8:00 A.M. to 7:00 P.M.   ¶GUIDED TOUR: 1 hour   ¶ON

Woodward Cave, Pennsylvania.

premises: snack bar, gift shop, camping, cabins, trailer camp, hunting and fishing, picnicking, sports area ¶NEARBY: restaurant, hotel ¶NEARBY AT-TRACTIONS: Penn's Cave, Poe Valley State Park, Boal Museum, Big Spring, Bald Eagle State Park, Reeds Gap State Park, Lincoln Caverns, Indian Caverns.

RHODE ISLAND

THERE ARE NO CAVES in the state of Rhode Island. One feature that is available for visitation is "Purgatory Chasm" in Newport. This small crevice in conglomerate has been enlarged by the waves.

SOUTH CAROLINA

SOUTH CAROLINA HAS A FEW CAVES in the central and western region but there are no caves open to the public.

SOUTH DAKOTA

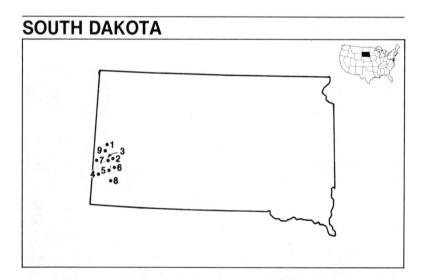

1 Bethlehem Cave 2 Black Hills Caverns 3 Diamond Crystal Cave 4 Jewel Cave National Monument 5 Rushmore Cave 6 Sitting Bull Crystal Caverns 7 Thunder Head Underground Falls (artificial cave) 8 Wind Cave National Park 9 Wonderland Cave

THE GREAT PLAINS underlie most of South Dakota. In the southwest corner of the state the Black Hills, a complex dome mountain, has extensive exposed limestones that are more than 300 million years old. They contain the greatest concentration of caves in the north central United States. The caves are noted for the profusion of large calcite crystals and delicate boxwork. Jewel Cave, one of the longest in the world, has more than 60 miles of mapped passageway.

BETHLEHEM CAVE

Mailing address: Bethlehem, SD 57708   *Phone:* (605) 787-4606   *Directions:* Take Exits 40 or 44 off Interstate 90 about 20 miles north of Rapid City, then west 3 miles.

Bethlehem Cave, South Dakota.

The natural entrance to Bethlehem Cave is 270 feet above the floor of Elk Canyon. Formerly known as Old Crystal Cave, it was reached via a narrow mine railway along the side of the canyon. Exhibited by owner Loui Storm it was only marginally successful. In 1952 Storm donated the property to the Benedictine Fathers. Since then Father Gilbert Stack, a monk of the Conception Abbey, and many volunteers have devoted their time to improving the cave and tending the Shrine of the Nativity located in the entrance room. Shown above is the statue of St. Benedict, the founder of the Benedictines and patron saint of cave explorers. Father Gilbert changed the name of the cave to Bethlehem Cave to honor both Jesus who was born in a stable in a cave, and St. Benedict who lived in a cave for three years before founding his first monastery.

It is a remarkable cavern more than three miles long of which a half mile is seen on the regular tour. The cave is lined with twelve-sided crystals of dogtooth spar ranging in size from ½ inch to 2 inches long. There is also popcorn, aragonite crystals, and cave pearls. The cave is easily reached by a road to the top of Elk Canyon. Father Gilbert, in keeping with the traditions of his order, works hard as a printer. The gift shop postmarks Christmas cards at the Bethlehem post office on the site; religious services are held in the cave on Christmas Eve in the Shrine of the Nativity.

¶OPEN: shrine open all year, 8:00 A.M. to 5:30 P.M.; lower cave open summer only   ¶GUIDED TOUR: 1 hour and 20 minutes   ¶ON PREMISES: gift shop, picnicking   ¶NEARBY: all facilities   ¶NEARBY ATTRACTIONS: Black Hills Caverns, Diamond Crystal Cave, Jewel Cave, Rushmore Cave, Sitting Bull Crystal Caverns, Thunder Head Underground Falls, Wind Cave, Mount Rushmore National Memorial, Reptile Gardens, Custer State Park, Castle Creek State Park, Deerfield Lake, Black Fox Recreation Area, Petrified Forest, Wonderland Cave.

BLACK HILLS CAVERNS

Mailing address: R.D. 8, Box 400, Rapid City, SD 57701   *Phone:* (605) 343-0542   *Directions:* 4 miles west of Rapid City on State Highway 44.

Black Hills Caverns was originally named Bennett's Wind Cave after a cowboy who noticed air moving in and out of the large natural entrance. In 1939 Dr. William Morse and his sons developed two levels and exhibited it as Wild Cat Cave. Features of the cave include the Roof Hole Room, named for the method of entry by early explorers, and formations of dogtooth spar, boxwork, and gypsum flowers. There are still undeveloped portions accessible only to energetic and specially equipped explorers.

¶OPEN: May 15 to September 15 from 8:00 A.M. to 8:00 P.M.   ¶GUIDED TOUR: 1 hour   ¶ON PREMISES: snack bar, gift shop, picnicking   ¶NEARBY: all facilities   ¶NEARBY ATTRACTIONS: Bethlehem Cave, Diamond Crystal Cave, Jewel Cave,

Rushmore Cave, Sitting Bull Crystal Caverns, Thunder Head Underground Falls, Wind Cave, Mount Rushmore National Memorial, Reptile Gardens, Custer State Park, Castle Creek State Park, Deerfield Lake, Black Fox Recreation Area, Petrified Forest, Wonderland Cave.

DIAMOND CRYSTAL CAVE

Mailing address: R.D. 8, Box 364, Rapid City, SD 57701   *Phone:* (605) 342-7162   *Directions:* About 3 miles west of Rapid City on State Highway 44.

The entrance to Diamond Crystal Cave is at the bottom of a canyon near Wild Irishman Gulch. Discovered in 1880, it was little visited until commercial development in 1929 when it was opened as Nameless Cave. The horizontal entrance leads to a series of rooms that are completely covered with calcite crystals of dogtooth spar. Some redissolving of the crystals has taken place, smoothing the surfaces of the sharp points, but many areas have excellent examples of the pristine translucent spar. In addition to popcorn and boxwork, the cave has some stalactites and stalagmites that were deposited after the water drained out following the formation of the crystals. There is an animal cave life exhibit and a display of petrified wood and minerals at the entrance building.

¶OPEN: May 1 to September 30 from 7:00 A.M. to 9:00 P.M.   ¶GUIDED TOUR: 40 minutes   ¶ON PREMISES: snack bar, gift shop, picnicking   ¶NEARBY: all facilities   ¶NEARBY ATTRACTIONS: Black Hills Caverns, Bethlehem Cave, Jewel Cave, Rushmore Cave, Sitting Bull Crystal Caverns, Thunder Head Underground Falls, Wind Cave, Mount Rushmore National Memorial, Reptile Gardens, Custer State Park, Castle Creek State Park, Deerfield Lake, Black Fox Recreation Area, Petrified Forest, Wonderland Cave.

JEWEL CAVE NATIONAL MONUMENT

Mailing address: Box 351, Custer, SD 57730   *Phone:* (605) 727-2301   *Directions:* About 14 miles west of Custer on U.S. 16.

A small whistling hole leading to a cave on the east side of Hell Canyon attracted the prospecting Michaud brothers in August of 1900. But their mining claim, called the Jewel Lode, was doubly disappointing. No valuable minerals were found, and their attempt to exhibit the cave as a tourist attraction failed. Although the cave was not a financial success, its natural formations prompted President Theodore Roosevelt to declare it a National Monument in 1908. The National Park Service assumed the responsibility for the area in 1934 and exhibited it to the public as a "small but interesting cavern." Tours were conducted by lantern up and down steep wooden, ladderlike stairs and along trails that were quite strenuous. In the 1950s some exploration was

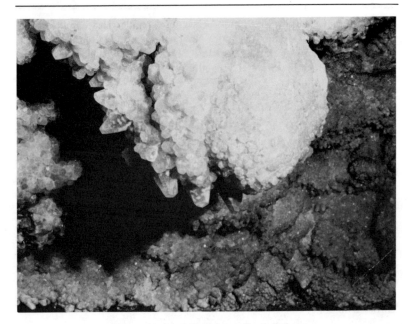

Jewel Cave National Monument, South Dakota.

permitted beyond the tourist trails; spelunkers attempted to follow the breeze whose whistle had initially attracted the Michaud brothers, and this elusive wind was to prove a tantalizing lure for a generation and more. Leaders in the search were Herb and Jan Conn, who have devoted more than 20 years to exploring, mapping, and seeking new avenues in this immense cave. Today there are more than 60 miles of mapped passageways making Jewel Cave the second longest in the United States. The Conns, a remarkable husband and wife team, discovered an area far more scenic than any previously known. Careful survey permitted the Park Service to sink an elevator shaft into the cave near Lithograph Canyon and exhibit this beautiful section. The new entrance permits access for further exploration as the "small but interesting" cave continues to grow.

There are three tours available, each providing a view of a different part of the cave and requiring a different degree of exertion. The Scenic (Modern) Tour is moderately strenuous. The visitor enters and leaves the cave by elevator from the Visitors' Center lobby. Some uphill walking and climbing long flights of stairs are necessary. The trail is paved and has aluminum stairs and handrails. The Historic (Primitive) Tour enters the cave through the natural entrance in Hell Canyon about 2 miles from the Visitors' Center. It is an unimproved trail and does not have electric lighting. Visitors are asked to carry candle lanterns. Old clothes are appropriate. The trail is quite strenuous and requires much climbing, bending, stooping, and sometimes crawling. The Spelunking Tour gives physically able persons a short introduction to the sport of caving. Sturdy boots are required. Old clothing should be worn as some of this trip will be on hands and knees, and some of it on one's stomach. (This brief experience will not qualify the participant as a cave explorer.) Advance registration for this tour is required.

¶OPEN: Visitors' Center open all year; guided tours May 15 to October 1, 8:00 A.M. to 7:30 P.M.   ¶GUIDED TOUR: There are three tours available: Scenic Tour, 1¼ hours; Historic Tour, 1½ hours; Spelunking Tour, 3 hours (advance registration required) ¶ON PREMISES: picnicking   ¶NEARBY: all facilities in Custer, 14 miles from cave   ¶NEARBY ATTRACTIONS: Bethlehem Cave, Black Hills Caverns, Diamond Crystal Cave, Rushmore Cave, Sitting Bull Crystal Caverns, Thunder Head Underground Falls, Wind Cave, Mount Rushmore National Memorial, Reptile Gardens, Custer State Park, Castle Creek State Park, Deerfield Lake, Black Fox Recreation Area, Petrified Forest, Wonderland Cave.

RUSHMORE CAVE

Mailing address: Keystone, SD 57751   *Phone:* (605) 255-4467   *Directions:* Near Mount Rushmore Memorial on State Highway 40 off U.S. 16A and Interstate 90.

The entrance to Rushmore Cave was discovered by placer miners in 1876 as they were constructing a flume for a gold mill in the valley below. There was no gold in the cavern, so the miners went about their business and left the cave for future explorers. In 1927 an artificial entrance was dug, trails and lights installed, and the cave exhibited to visitors who were attracted to the area by the nearby Mount Rushmore Memorial. The tour consists of a series of rooms connected by joint-controlled passages terminating in a large, well-decorated room. This is the only cave open to the public in the Black Hills that does not have any of the crystals or dogtooth spar common to the other caves of the region. The decorations seen include numerous stalactites and helictites. Portions of the cave are undeveloped and it is possible that there is more passageway to be discovered.

¶OPEN: May through October 8:00 A.M. to 6:00 P.M.; June, July and August from 7:00 A.M. to 9:00 P.M.   ¶GUIDED TOUR: 1 hour   ¶ON PREMISES: snack bar, gift shop   ¶NEARBY: restaurant, camping, motels, hotel, cabins, trailer camp, picnicking   ¶NEARBY ATTRACTIONS: Bethlehem Cave, Black Hills Caverns, Diamond Crystal Cave, Jewel Cave, Sitting Bull Crystal Caverns, Thunder Head Underground Falls, Wind Cave, Mount Rushmore National Memorial, Reptile Gardens, Custer State Park, Castle Creek State Park, Deerfield Lake, Black Fox Recreation Area, Rushmore Tramway, Wonderland Cave.

Rushmore Cave, South Dakota.

SITTING BULL CRYSTAL CAVERNS

Mailing address: Rapid City, SD 57702   *Phone:* (605) 342-2777   *Directions:* About 9 miles south of Rapid City on U.S. 16.

The natural entrance to Sitting Bull Crystal Caverns is located near the bottom of Rockerville Gulch where a masonry staircase leads down a steeply sloping crevice to the main cave about 100 feet below. Although it was probably known by the Indians, the first report of the cave was from prospectors in 1876 who scoured the hills during the mini-gold rush of the region. Opened in 1930, the cavern displays several rooms that have the finest dogtooth spar in the Black Hills. The crystal is confined to areas resembling geodes that bristle with the spar. It has formed in four different layers in some rooms, and there are some individual crystals that exceed 16 inches in length. Sitting Bull Crystal Caverns is a small cave in volume, but the fine display of crystal makes it an important one to visit.

Sitting Bull Crystal Caverns, South Dakota.

¶OPEN: May 15 to December 15 from 7:00 A.M. to 7:30 P.M.   ¶GUIDED TOUR: 30 minutes   ¶ON PREMISES: gift shop, picnicking   ¶NEARBY: all facilities   ¶NEARBY ATTRACTIONS: Bethlehem Cave, Black Hills Caverns, Diamond Crystal Cave, Jewel Cave, Rushmore Cave, Thunder Head Underground Falls, Wind Cave, Mount Rushmore National Memorial, Reptile Gardens, Custer State Park, Rushmore Tramway, Wonderland Cave.

THUNDER HEAD UNDERGROUND FALLS (artificial cave)

Mailing address: Rapid City, SD 57702   *Phone:* (605) 343-0081   *Directions:* On State Highway 44 about 10 miles west of Rapid City.

This is not a natural cave but a bypass tunnel built by miners in the 1880s in an attempt to drain about one and a half miles of Rapid Creek at Big Bend. The miners planned a shaft to cut through the narrow neck of the meander loop that would drain the creek and permit them to search for gold. Tunnels were started on each side of the loop; one sloped up, the other down, and they met in the middle. The connection was made, but an error in elevation caused a waterfall when the tunnel was used as a drain. The project succeeded in draining Big Bend, but the gold did not materialize. The tunnel was abandoned and the project forgotten for 75 years. In 1950, Albert and Vera Eklund decided that the waterfall in the tunnel made a spectacular scene. They made the necessary improvements and opened the tunnel as an attraction for summer visitors. The falls flow at a steady rate of eight cubic feet per second and provide a thundering spectacle.

¶OPEN: April 1 to October 31 during daylight hours   ¶SELF-GUIDED TOUR: 30 minutes; no lights required   ¶ON PREMISES: no facilities   ¶NEARBY: all facilities   ¶NEARBY ATTRACTIONS: Black Hills Caverns, Diamond Crystal Cave, Jewel Cave, Rushmore Cave, Sitting Bull Crystal Caverns, Wind Cave, Mount Rushmore National Memorial, Reptile Gardens, Custer State Park, State Fish Hatchery, Pactola Dam, Canyon Lake, Wonderland Cave.

WIND CAVE NATIONAL PARK

Mailing address: Hot Springs, SD 57747   *Phone:* (605) 727-2301   *Directions:* On U.S. 385, 12 miles north of Hot Springs.

On the southeastern slopes of South Dakota's Black Hills, Wind Cave National Park provides an opportunity to look back in time. Above ground is the unspoiled prairie as the settlers saw it, with free-roaming bison, antelope, and prairie dogs, all without fences. Beneath this bucolic scene lies a maze of subterranean passages and rooms that take the visitor back millions of years into geologic time.

Tom Bingham, a Black Hills pioneer, discovered the cave in 1881 while deer hunting. He was attracted by the whistling of the wind through a small hole in the rock. In 1890 mining claims were filed and the first serious explorations were made by the McDonald family, who later formed a company and operated the cave as a visitor attraction. In 1903 President Theodore Roosevelt signed a bill establishing Wind Cave National Park. More than 27 miles of passageway are known, and exploration still continues. About one and a half miles are open to the public, including excellent examples of boxwork, popcorn, and gypsum flowers. The original entrance has been bypassed by an artificial tunnel with air lock doors to prevent significant loss of air, so the "wind" effect of the original entrance can still be felt.

¶OPEN: all year 8:00 A.M. to 5:00 P.M. with extended hours in summer ¶GUIDED TOUR: There are three tours available in summer; short tour about 1 hour; historical candlelight tour 1¾ hours; spelunking tour about 3 hours (requires advance registration). All tours enter on foot and exit by elevator, 500 to 700 steps depending upon the tour selected. The spelunking tour gives physically able persons a short introduction to the sport of caving. Sturdy, ankle high, lace boots are required. Old clothing should be worn as some crawling is necessary. Those who take this tour will not be qualified as cave explorers as a result of their brief experience   ¶ON PREMISES: (summer only) restaurant, snack bar, gift shop, camping, trailer camp. Picnicking all year ¶NEARBY: all facilities in Hot Springs 12 miles away   ¶NEARBY ATTRACTIONS: Bethlehem Cave, Black Hills Caverns, Diamond Crystal Cave, Jewel Cave, Rushmore Cave, Sitting Bull Crystal Caverns, Thunder Head Underground Falls, Mount Rushmore National Memorial, Reptile Gardens, Custer State Park, Wonderland Cave.

WONDERLAND CAVE

Mailing address: Box 83, Nemo, SD 57759   *Phone:* (605) 343-5043   *Directions:* Off Interstate 90 and State Highway 14 between Rapid City and Sturgis, or off U.S. 385 via Nemo.

The Wonderland Cave entrance is located high on the side of a canyon about a mile south of Bethlehem Cave. It is an extensive cave on two distinct levels and features many stalactites and stalagmites as well as dogtooth spar and helictites. There is no indication that it was known before 1929. The development in 1930 has protected the delicate formations and features, presenting them in a condition that is pristine and natural.

¶OPEN: daily May 15 to October 1, 7:00 A.M. to 10:00 P.M.¶GUIDED TOUR: 45 minutes to 1 hour   ¶ON PREMISES: snack bar, gift shop, picnicking, hiking and cycle trails   ¶NEARBY: restaurants, camping, motel, hotel, cabins, trailer camp   ¶NEARBY ATTRACTIONS: Bethlehem Cave, Black Hills Caverns,

Diamond Crystal Cave, Jewel Cave, Rushmore Cave, Sitting Bull Crystal Caverns, Thunder Head Underground Falls, Wind Cave, Castle Creek State Park, Black Fox Recreation Area.

TENNESSEE

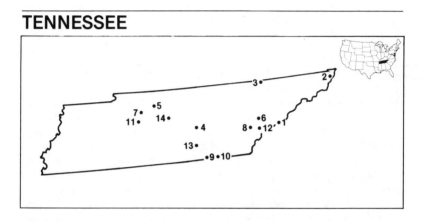

1 Alum Cave 2 Bristol Caverns 3 Cudjo's Cave 4 Cumberland Caverns 5 Cumberland Museum and Science Center (artificial caves) 6 Forbidden Caverns 7 Jewel Cave 8 Lost Sea 9 Racoon Mountain Caverns 10 Ruby Falls (Lookout Mountain Caverns) 11 Ruskin Cave 12 Tuchaleechee Caverns 13 Wonder Cave

THE CUMBERLAND PLATEAU on the west slope of the Appalachian Mountains bifurcates the state of Tennessee with the Highlands to the east and the Coastal Plain to the west. This plateau, an extension of the great limestone region of Kentucky, has some of the finest caves in the eastern United States. There are hundreds of wild caves and a dozen extensive show caves with outstanding features. Saltpeter vats, Indian artifacts, gypsum flowers, waterfalls, and prime examples of cave scenery are found in the caves of this region.

ALUM CAVE

Mailing address: Great Smoky Mountains National Park, Gatlinburg, TN 37738   *Phone:* (615) 436-5615   *Directions:* On U.S. 441 in Great Smoky National Park. Foot trail to cave, about 5½ miles.

This is not a true cave, but a large overhanging bluff along what is known as Alum Cave Trail. Information can be obtained from Park Headquarters or any Park Ranger. The trail is well marked and maps of the region are available showing the location of the cave.

BRISTOL CAVERNS

Mailing address: 958 West State Street, Bristol, TN 37620   *Phone:* (615) 764-1646   *Directions:* On U.S. 421 about 5 miles east of Bristol.

One entrance to Bristol Caverns is a skylight opening at the top of a huge gallery where, according to legend, a grisly murder was committed and the opening used to dispose of the body. Another entrance a half mile away, the resurgence of Sinking Creek, was used by Indians as a hiding place after raids on the early settlers.

The visitors' entrance was discovered by excavations at a spring near the top of a ridge which permits access to one of the most highly decorated rooms in the cave. The tour descends to the bed of Sinking Creek, the lowest point, and then returns to the surface. The entire trip takes place in one huge chamber more than 2500 feet long and averaging 60 feet wide. The formations and rock outcroppings provide partitions that make the tour seem a series of separate rooms. A further extension, not open to the public, leads to one of the largest formations in the country. This large and beautiful cave should appeal to any visitor.

¶OPEN: all year. Summer 8:00 A.M. to 8:00 P.M.; winter 8:00 A.M. to 6:00 P.M. except Sundays from 12 noon to 6:00 P.M.   ¶GUIDED TOUR: 1 hour   ¶ON PREMISES: snack bar, gift shop, camping, picnicking   ¶NEARBY: all facilities ¶NEARBY ATTRACTIONS: South Holston Lake, Boone Lake, Barter Theater, Warrior's Path State Park, Natural Tunnel (Virginia), Cudjo's Cave (about 40 miles).

CUDJO'S CAVE

Mailing address: Cumberland Gap, TN 37724   *Phone:* (703) 861-2203   *Directions:* On U.S. 25E in Cumberland Gap National Historical Park.

Cudjo's Cave is located almost exactly at the tricorner of Tennessee, Virginia, and Kentucky. The entrance is in Lee County, Virginia, but access is from Tennessee and reportedly one corner of the Castle Room is in Kentucky. The entrance and exit of the cave is about halfway up the side of Pinnacle Mountain in Cumberland Gap National Historical Park. Discovered in 1750 by Dr. Thomas Walker, the cave was well known locally but little noticed elsewhere. The name Cudjo's Cave was used by J. T. Trowbridge as the title of a romantic Civil War novel published in 1863; the success of that story made the name well known nationally, but there is no evidence that the author had ever visited the cave. There are five levels. The entrance is on level number 4 with stairways leading up and down. Features include large well-decorated rooms, a moonshine still, and artifacts dating back to Civil War days. Although the cave is located within a National Park it is operated under an agreement with Lincoln Memorial University.

¶OPEN: all year 9:00 A.M. to 5:00 P.M.; extended hours from Memorial Day to Labor Day   ¶GUIDED TOUR: 45 minutes   ¶ON PREMISES: gift shop ¶NEARBY: restaurant, camping, trailer camp, motel, hotel, picnicking ¶NEARBY ATTRACTIONS: Fern Lake, Pine Mountain State Park (Kentucky), Big Ridge State Park, Norris Lake, Norris Dam, Cherokee Lake, Natural Tunnel (Virginia).

CUMBERLAND CAVERNS

Mailing address: McMinnville, TN 37110   *Phone:* (615) 668-4396   *Directions:* 6 miles southeast of McMinnville, off State Highway 8.

Cumberland Caverns, Tennessee. Photo by Roy Davis.

Under Cardwell Mountain lies Cumberland Caverns, the most extensive cave system known in the state of Tennessee. Exploration has revealed more than 23 miles of passageway and several entrances. Discovered in 1810 by a surveyor, Aaron Higgenbotham, it was called Higgenbotham Cave for nearly 150 years. It has been used as a mine for saltpeter, a picnic area, and a place to explore for local adventurers. In 1950 systematic and serious exploration in the remote portions of the cave by enthusiastic and energetic explorers resulted in the establishment of a corporation to make the cave accessible for visitors. One of the leaders, Roy Davis, has developed and designed the lighting in this as well as in more than 30 other caves throughout the United States.

The Henshaw entrance seen on the tour was used for saltpeter mining during the Civil War. Some of the original vats are shown, along with a more recent moonshine still. Visitors also see the Ten Acre Room, a corridor 60 feet wide and 1400 feet long where early explorers camped while exploring deeper into the cave. Other features include the Mountain Room, a huge unsupported ceiling 210 feet wide and more than 500 feet long, and a spectacular display of large formations still active and growing. Only a small part of this large cave can be shown. Not seen on the regular tour are outstanding displays of gypsum flowers, selenite needles, and moon milk. Cumberland Caverns was the first to provide an underground overnight camping experience for Boy Scouts and continues to be the most popular.

¶OPEN: May through October 9:00 A.M. to 5:00 P.M. daily June 1 to August 31; weekends only May, September and October. By appointment only November through April   ¶GUIDED TOUR: 1½ hours. Special groups for overnight camping in the cave by reservation   ¶ON PREMISES: snack bar, gift shop, camping, trailer camp, picnicking, underground dining room for special parties ¶NEARBY: restaurant, motel, hotel, cabins   ¶NEARBY ATTRACTIONS: Fall Creek State Park, Center Hill Reservoir, Rock Island Lakes, Cumberland Mountain State Park, Wonder Cave.

CUMBERLAND MUSEUM AND SCIENCE CENTER (artificial caves)

Mailing address: 800 Ridley Avenue, Nashville, TN 37203   *Phone:* (615) 242-1858   *Directions:* Take Exit 210C off Interstate 240 and follow signs to Museum Center. Cave exhibits are on the second floor.

There are three small exhibits in this museum providing habitat displays of animals that live in caves. The first is an ice cave with polar bears; the second an underwater cave showing a coral reef and assorted tropical water fauna; the third displays a cave scene with formations and bats to show the conditions that might be found in such a habitat. This exhibit is not designed as a cave display, wildlife being the principal interest.

¶OPEN: all year from 10:00 A.M. to 5:00 P.M. Tuesday through Saturday; Sunday 1:00 P.M. to 5:00 P.M.; closed on Mondays   ¶ON PREMISES: snack bar, gift shop   ¶NEARBY: all facilities.

FORBIDDEN CAVERNS

Mailing address: Route 8, Sevierville, TN 37862   *Phone:* (615) 453-5972
Directions: About 1 mile off U.S. 411 between Sevierville and Newport.

Forbidden Caverns' history dates back to Cherokee days, and includes a romantic legend of an Indian princess "sealed in a hollow mountain." Located on the northwestern slope of English Mountain, it was first reported in 1919, and known as Blowing Cave. In the 1930s it housed an illegal moonshine still that used the ample water supply of the underground river. A more prosaic use was made later when limestone was quarried from the entrance area. In 1967 it was opened to the public, and trails were built to permit access to the stream passage and more remote rooms. Along the stream passage both solution and erosion are shown in the meandering channels.

¶OPEN: April 1 to October 31 daily 10:00 A.M. to 5:00 P.M. except June 15 through Labor Day 9:00 A.M. to 6:00 P.M.   ¶GUIDED TOUR: 1 hour   ¶ON

Forbidden Caverns, Tennessee.

premises: snack bar, gift shop, picnicking   ¶NEARBY: camping, restaurant, trailer camp, motel, hotel, cabins   ¶NEARBY ATTRACTIONS: Cherokee Lake, Gatlinburg Resort Area, Douglas Lake, Lost Sea, Tuckaleechee Caverns, Sam Houston Schoolhouse.

JEWEL CAVE

Mailing address: Dickson, TN 37055   *Phone:* (615) 763-9144   *Directions:* Off U.S. 70 north on State Highway 46 about 10 miles.

Entrance to Jewel Cave is from the side of a small hill a few feet from Yellow Creek Road. Discovered in 1885 by workmen quarrying limestone, it was first explored by the young daughters of Tom Rogers as they crawled through the water of a spring into the main chamber. Exploration disclosed only one room, but the beautiful formations found there made the cave a popular local visitation spot. In 1925 it was opened to visitors and during construction a second room was discovered, doubling the size of the cave. An artificial entrance was cut to permit easy access. Completion of the development provided one of the easiest and most attractive cave tours in Tennessee. In addition to the fine display of formations, the excavations disclosed a number of Pleistocene bones

Jewel Cave, Tennessee.

of animals that lived in the area more than 25,000 years ago. These artifacts are on display in the cave and entrance building.

¶OPEN: Memorial Day to Labor Day 9:00 A.M. to 5:00 P.M.   ¶GUIDED TOUR: 1 hour   ¶ON PREMISES: snack bar, gift shop, picnicking   ¶NEARBY: restaurant, motel, camping, trailer park   ¶NEARBY ATTRACTIONS: Ruskin Cave, Jefferson Davis Monument, Donelson National Military Park and Cemetery, Montgomery Bell State Park.

LOST SEA

Mailing address: Route 2, Lost Sea Pike, Sweetwater, TN 37874   *Phone:* (615) 337-6616   *Directions:* Between Sweetwater and Madisonville on U.S. 68.

The natural entrance to this cave, a small opening in the side of a ridge, was discovered by the Indian Chief Craighead. White settlers stored potatoes here, and during the Civil War used the cave to manufacture saltpeter. In 1905 exploration in what was then known as Craighead Caverns by Ben Sands revealed the Lake Room and the Lost Sea. An artificial entrance was dug in 1927 permitting access to a 100-foot-wide passage 30 feet high. Lighting, walkways, and an underground dance floor were installed, but the Lost Sea was not exhibited because of the difficulty of access. The cave business did not prosper,

The Lost Sea, Tennessee.

and by 1940 the cave was being used as a mushroom farm. However, explorers discovered the skeleton of an extinct jaguar that had perished underground more than 25,000 years ago. In 1963 the cave was reopened as Lost Sea, and featured a boat ride on the huge underground lake. The present tour also shows the early saltpeter workings, the mushroom farm, the dance floor, and a display of anthodite flowers.

¶OPEN: daily all year 9:00 A.M. till sundown   ¶GUIDED TOUR: 55 minutes   ¶ON PREMISES: restaurant, gift shop, picnicking   ¶NEARBY: camping, motel, hotel, trailer camp   ¶NEARBY ATTRACTIONS: Forbidden Caverns, Oxone Falls, Great Smoky National Park, Tuckaleechee Caverns, Watts Bar Lake, Fort Loudoun, Quinn Springs State Park, Chilhowee State Park

RACOON MOUNTAIN CAVERNS

Mailing address: Route 4 Cummings Highway, Chattanooga, TN 37409 *Phone:* (615) 821-9403   *Directions:* On U.S. 41 off Interstate 24 about 5 miles west of Chattanooga.

Racoon Mountain Caverns, originally developed in 1931 as Tennessee Caverns and later called Crystal Caverns, is located at the base of Racoon Mountain. Entrance is through the lobby of a building built into the side of the hill directly into a large room that is well decorated and more than a hundred feet in diameter. The tour extends through several corridors and without retracing steps returns to the entrance room via an artificial tunnel. An extensive undeveloped area beyond the tourist route is used in part for a Scout program during the winter months. Special guided tours are arranged to provide a spelunking experience to young people equipped with flashlights and helmets. In addition to the cave, a cable car skyride ascends to the top of Racoon Mountain, 2800 feet above the entrance building. Hang-gliding enthusiasts use this as a launching spot for gliders into the Lookout Mountain valley below.

¶OPEN: all year except Thanksgiving, Christmas and New Year's Day. Summer 9:00 A.M. to 9:00 P.M.; winter 9:00 A.M. to 5:30 P.M.   ¶GUIDED TOUR: ½ hour   ¶ON PREMISES: restaurant, gift shop, camping, trailer camp, picnicking, hang gliding in summer   ¶NEARBY: all facilities in Chattanooga   ¶NEARBY ATTRACTIONS: Rock City and Gardens, Ruby Falls, Lookout Mountain Incline Railway, Confederama, Russell Cave National Monument (Alabama), Wonder Cave, Booker T. Washington State Park, Signal Mountain.

RUBY FALLS (LOOKOUT MOUNTAIN CAVERNS)

Mailing address: P.O. Box 3160, Chattanooga, TN 37404   *Phone:* (615) 821-2544   *Directions:* On State Highway 148 just 3 miles from downtown Chattanooga atop Lookout Mountain.

On the side of Lookout Mountain overlooking downtown Chattanooga and Moccasin Bend of the Tennessee River is a stone building at the entrance to Ruby Falls. Originally this site was to be the access to Lookout Mountain Caverns, some 420 feet below. But during excavation of the elevator shaft in 1923, another cave was discovered at the 260-foot level. The new one had a natural waterfall which cascaded from the ceiling into a pool in the center of a huge bell-shaped room. It was named Ruby Falls for the wife of Leo Lambert,

Ruby Falls (Lookout Mountain Caverns), Tennessee.

an early promoter and explorer. At first the two caves were exhibited on separate tours but the popularity of the falls far exceeded that of the lower cave, and that trip was discontinued. The falls are heard roaring and whistling before they are reached. In the waterfall room, winds stirred up by the falling water sweep over the observers. The narrow and winding passage from the elevator to the falls is the natural drainage route for the water and offers an interesting contrast to the impressive waterfall.

¶OPEN: daily all year 7:30 A.M. to 8:00 P.M.   ¶GUIDED TOUR: 1 hour   ¶ON PREMISES: snack bar, gift shop, elevator   ¶NEARBY: all facilities   ¶NEARBY ATTRACTIONS: Rock City Gardens, Russell Cave National Monument (Alabama), Crystal Cave, Wonder Cave, Harrison Bay State Park, Booker T. Washington State Park, Signal Mountain.

RUSKIN CAVE

Mailing address: Dickson, TN 37055   *Phone:* (615) 763-9141   *Directions:* Off U.S. 70 on State Highway 46 (Yellow Creek Road) about 10 miles north.

The impressive entrance of Ruskin Cave in Yellow Creek Valley has been a landmark since early Indian times. John Ruskin was an English author, art critic, and social theorist who started a communal colony here. The colony, which included shops, mills, and factories, was successful for only a few years, disbanding in 1899. The property became a site for Ruskin Cave College, and later the cave was a dance hall and meeting place. Today the floor of the entrance room is paved to permit shows and square dancing. Beyond this point it continues as an unlighted passage about four feet high extending a few hundred feet back into the cave. Tracks of the cart used by the Ruskinites to transport vegetables into the cool interior can be seen, but there are no formations or decorations. The cave is only one feature of a family resort with facilities for swimming, hiking, and other sports.

¶OPEN: daily Memorial Day to Labor Day 9:00 A.M. to 5:00 P.M.   ¶SELF-GUIDED TOUR: except in entrance room. Bring your own lights for further exploration.   ¶ON PREMISES: restaurant, snack bar, gift shop, camping, cabins, trailer camp, picnicking, swimming pool, recreational area, Haunted House exhibit   ¶NEARBY: motels, hotel   ¶NEARBY ATTRACTIONS: Jewel Cave, Jefferson Davis Monument, Donelson National Military Park and Cemetery, Kentucky Lake, fishing and swimming.

TUCKALEECHEE CAVERNS

Route 1, Townsend, TN 37882   *Phone:* (615) 448-2274   *Directions:* Off State Highway 73 between Maryville and Gatlinburg.

At the foot of Little Mountain in Dry Valley lies the natural entrance to Tuckaleechee Caverns, which was discovered by lumbermen. Their original explorations using pine torches and kerosene lanterns were limited to the stream passage. In 1953, Bill Vananda and Harry Myers pooled their resources and through their own hand labor opened the cave to visitors. The first tours were by Coleman lanterns, but later electric lights and concrete trails were added. Today Tuckaleechee Caverns, with a half-mile of walkway, is noted for tall

Tuckaleechee Caverns, Tennessee.

stalagmites called totem poles, broomsticks, and numerous soda straw formations. Visitors also see the largest room in the cave, 100 feet by 160 feet by 65 feet high, and the three independent streams that merge into the main stream within the cave. It is believed that the water emerges about ¾ of a mile away at Dunn Spring.

¶OPEN: daily all year 9:00 A.M. to 6:00 P.M.; summer open to 7:00 P.M. ¶GUIDED TOUR: 1 hour ¶ON PREMISES: gift shop, camping, picnicking ¶NEARBY: snack bar, restaurant, motel, hotel, cabins, trailer camp ¶NEARBY ATTRACTIONS: Lost Sea, Forbidden Caverns, Gatlinburg resort area, Smoky Sky Lift, Great Smoky National Park, Douglas Lake, Fontana Lake and Dam.

WONDER CAVE

Mailing address: Monteagle, TN 37356 *Phone:* (615) 467-3540 *Directions:* About 2 miles north of the Monteagle Exit on Interstate 24.

The natural entrance of Wonder Cave on the north slope of Layne Cove is in a spring that has been enlarged to permit access to a corridor formed by the underground stream. Discovered in 1897, it was opened to the public the next year, and lighted with acetylene gas from a generator at the entrance. The cave consists primarily of a main corridor which exceeds a mile in length in a nearly straight line. There are numerous formations in a discontinuous upper level. The tour follows the stream bed past sculptured walls cut by the flowing water. The cave is exhibited by gasoline lantern.

Wonder Cave, Tennessee.

¶OPEN: all year 7:00 A.M. to dark   ¶GUIDED TOUR: 1 hour   ¶ON PREMISES: gift shop, camping, trailer camp, picnicking, antique shop   ¶NEARBY: restaurant, snack bar, motels, hotels, cabins   ¶NEARBY ATTRACTIONS: Cumberland Caverns, Ruby Falls, Sequoyah Caverns (Alabama), Crystal Cave, Rock City and Gardens, Lookout Mountain, Signal Mountain.

TEXAS

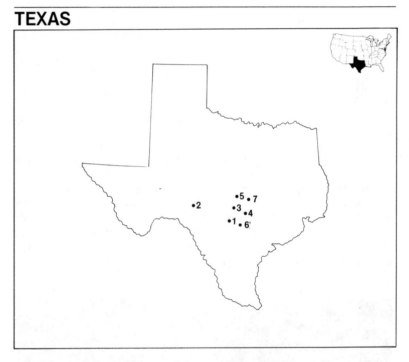

1 Cascade Caverns 2 Caverns of Sonora 3 Cave-Without-A-Name 4 Inner Space 5 Longhorn Cavern State Park 6 Natural Bridge Caverns 7 Wonder World

SPRAWLING TEXAS IS SPREAD over three distinct landforms: to the east are the Gulf Coastal Plains; in the center, the lower Western Plains; and in the extreme west the extension of the Guadalupe Mountains. There are hundreds, perhaps thousands, of caves in this huge and geologically complex area, but only a few are open to the public. Most are found clustered around the populated areas of central Texas. The Balcones Escarpment near Austin and San Antonio outlines the edge of the Edwards Plateau, a limestone block that contains some of the largest caves in the state. There is little of historical interest in the show caves — some Indian artifacts and some bones. Most of the show caves have been recently discovered by drilling and quarrying and show no history of man.

CASCADE CAVERNS

Mailing address: Route 1, Box 57A, Boerne, TX 78006   *Phone:* (512) 755-9285   *Directions:* About 16 miles northwest of San Antonio on Interstate 10 at Boerne, then 5 miles east on State Highway 46.

The hill country of central Texas contains hundreds of fine caves, some of them open to the public for tours; others are accessible only with proper spelunking

Cascade Cavern, Texas.

equipment. Cascade Caverns offers both in its program, providing an opportunity for the visitor to take a conventional guided tour in street clothes or by special arrangement to go down to the lower levels for a look at the undeveloped section. The natural entrance has for years been the catchment area for "gully washers," heavy storm water that overflows the natural surface drainage of the area. This condition has been corrected by diverting the water and by constructing dams upstream of the cave. The visitor can see traces of the flowing stream in the eroded formations and scoured walls of the cave passages. This is not a recent occurrence, for deep in the cave are several mastadon bones still embedded in the clay. The end of the walking tour is at the Cascade Room, where a fine waterfall pours out of a crevice at the ceiling and splashes down into a pool on the floor.

The wild cave trip gains access to the lower cave through a 24-inch vertical standpipe that has been installed to control the level of the pool. Only the hardy and energetic should tackle that route. The tour of the lower cave shows a continuation of the ancient watercourse, now partially flooded. It provides an exciting trip. Written, prior permission must be obtained to enter the undeveloped portions of the cave. The regular visitor tour, however, gives a fine picture of the kind of features you would see on the lower route.

¶OPEN: all year 9:00 A.M. to 5:00 P.M.   ¶GUIDED TOUR: about 50 minutes. Tour to the lower portion of the cave requires special equipment and lights. Contact Mr. John Bridges for reservations and times of tours   ¶ON PREMISES: snack bar, gift shop, camping, trailer camp, picnicking, swimming pool, dance pavilion   ¶NEARBY: motel, cabins, hotel, restaurant   ¶NEARBY ATTRACTIONS: Natural Bridge Caverns, Cave-Without-A-Name, San Antonio Missions, Wonder Cave, Inner Space.

CAVERNS OF SONORA

Mailing address: P.O. Box 213, Sonora, TX 76950   *Phone:* (915) 387-3105 *Directions:* Off Interstate 10 about 8 miles west of Sonora, take Caverns of Sonora Road south, about 3 miles.

The Caverns of Sonora are exceptional among caves open to the public in the United States by virtue of the beauty, variety, and accessibility of its speleothems. The entrance rooms, drab and mundane, were known for many years by local ranchers as Mayfield Cave; little interest was taken in it until several spelunkers explored the deeper recesses in 1955. Their discovery brought attention to the cave, and in 1960 it was opened to the public with trails and electric lights. The development by Jim Papadakis, Jack Burch and Stanley Mayfield opened to view one of the most unusual caves in the world.

The visitor is treated to a walk through rooms lined with crystals, helictites, soda straws and delicate formations that are breathtakingly beautiful. No other cave offers so many formations at such close proximity to the visitor.

Caverns of Sonora, Texas.

This proliferation of exotics has invited some vandalism by tourists. This desecration, unthinkable to any reasonable person, nevertheless exists and unless extraordinary care is taken to protect the undeveloped parts of the cave, future generations may never have the opportunity to see what might be a unique display in the world.

¶OPEN: all year 8:00 A.M. to 6:00 P.M.   ¶GUIDED TOUR: 1½ hours. Tours leave every 30 minutes.   ¶ON PREMISES: snack bar, gift shop, camping, trailer camp, picnicking ¶NEARBY: restaurant, motel ¶NEARBY ATTRACTIONS: ranches.

CAVE-WITHOUT-A-NAME

Mailing address: Route 2, Box 99B, Boerne, TX 78006 *Phone:* (512) 537-4212 *Directions:* From Boerne take County Route 474 north about 11 miles.

Cave-Without-A-Name is unfortunately almost a cave without a road—it is reached by a five-mile rocky route. The result has been poor visitation of what many regard as one of the prettiest caves in Texas. It is the second longest cave in the state, with nearly five miles of passage, and the portion visited exhibits about 1000 feet of splendid cave scenery. A masonry staircase descends about 70 feet to the main corridor. The trail is then level and smooth for the remainder of the trip. The tour ends with a tantalizing view of a river passage with a high vaulted roof that disappears into the darkness. There is no brochure available describing this cave, which is listed on some road maps as Century Caverns. The facilities at the entrance are not encouraging, but it is a fine cavern well worth visiting.

¶OPEN: on weekends only 8:00 A.M. to 6:00 P.M. ¶GUIDED TOUR: 45 minutes ¶ON PREMISES: gift shop, picnicking, camping ¶NEARBY: restaurants, motel and other facilities in Boerne (11 miles) ¶NEARBY ATTRACTIONS: Cascade Caverns, Natural Bridge Caverns, Wonder Cave, Blanco State Park, Kerrville State Park, The Alamo, Fort Sam Houston, Landa Park, Aquarena, New Braunfels.

INNER SPACE

Mailing address: Box 451, Georgetown, TX 78626 *Phone:* (512) 863-5545 *Directions:* On Interstate 35 about 2 miles south of Georgetown.

Men of the Texas Highway Department discovered this cave in 1963 while drilling a test hole for supports for a highway overpass on Interstate 35 between Austin and Georgetown. As they were drilling, the bit suddenly dropped 26 feet into a void. The hole was enlarged to 24 inches in diameter to permit human entry. Speleologists entered and mapped 1½ miles of remarkable virgin cave, returning to the surface with wonderful tales of rooms and scenery below. A local corporation developed the cavern and opened it to the public in 1965. A notable feature of the cave is its display of fossil bones of peccaries and mammoth tusks, some more than 20,000 years old. These animals were trapped in open sinkholes that have long since filled with debris. There is no evidence that man was ever in the cave before the drill bit broke through the ceiling. Inner Space, within sight of the Interstate, is one of the most conveniently located caverns in the United States. A cable car transports visitors into the cave, and nearly level trails make the trip easy and comfortable. The cave is as pristine and unspoiled as it was when speleologists first explored it.

Inner Space, Texas.

¶OPEN: daily all year from 10:00 A.M. to 5:00 P.M.   ¶GUIDED TOUR: 1 hour and 15 minutes   ¶ON PREMISES: snack bar, gift shop, picnicking, cable car, country store   ¶NEARBY: restaurant, motel, cabins, hotel, camping, trailer camp   ¶NEARBY ATTRACTIONS: Natural Bridge Caverns, Wonder Cave, Aquarena, Cave-Without-A-Name, Cascade Caverns, Longhorn Cavern, Highland Lakes Area.

LONGHORN CAVERN STATE PARK

Mailing address: Burnet, TX 78611   *Phone:* (512) 756-4680   *Directions:* Off U.S. 281 on Park Road about 6 miles west.

The entrance to Longhorn Cavern in the hill country of central Texas is a collapsed sink in the limestone. A natural bridge, a remnant of the roof, spans the path leading to the cave. Once a retreat for Indians, the Main Room was used to store gunpowder for the Confederates during the Civil War. Sam Bass, notorious bandit of 1870s, utilized the cave for a hideout. In 1932 two miles of lighted trails were opened to the public by the state of Texas. The cave's convoluted walls, dramatically sculptured by water, show the meanders of the stream that formed them.

Longhorn Cavern, Texas.

¶OPEN: daily all year from 10:00 A.M. to 3:00 P.M. except from June through August open until 5:00 P.M. From October to February closed on Monday and Tuesday   ¶GUIDED TOUR: 2 hours   ¶ON PREMISES: snack bar, gift shop, picnicking   ¶NEARBY: motel, restaurant, camping, trailer camp   ¶NEARBY AT-TRACTIONS: Highland Lakes Area, Inks Lake State Park, Inner Space, Wonder Cave, Cave-Without-A-Name, Cascade Caverns, Natural Bridge Caverns.

NATURAL BRIDGE CAVERNS

Mailing address: Route 3, Box 515, Natural Bridge Caverns, TX 78218 *Phone:* (512) 651-6101 *Directions:* Between San Antonio and New Braunfels on Natural Bridge Caverns Road Exit. County Route 3009 about 5 miles west.

Natural Bridge Caverns, Texas.

Wonder World, Texas.

Natural Bridge Caverns is the largest show cave in Texas and is located in a beautiful section of the Texas hill country. The cave was discovered in 1960 by a group of college student spelunkers. The first to enter the cave was Orion Knox Jr., who later aided in the construction of the tourist route. Several years of strenuous work by Jack Burch, Harry Heidemann, Orion Knox and others have provided a sensitive and tasteful presentation of this fine cave. A sloping trail descends into the entrance beneath a limestone natural bridge that gives the cave its name. An easy tour of nearly a mile takes visitors into beautifully lighted rooms that have huge formations and spectacular vistas which appear to be as pristine as when they were first discovered. To preserve the moisture within the cave, glass doors have been installed on the entrance and exit tunnels. This precaution has prevented the drying out of the formations and most are still actively growing.

¶OPEN: all year summer 10:00 A.M. to 6:00 P.M.; winter 10:00 A.M. to 4:00 P.M. ¶GUIDED TOUR: 1¼ hours   ¶ON PREMISES: snack bar, gift shop, picnicking ¶NEARBY: camping, trailer camp, restaurant, motel, cabins   ¶NEARBY ATTRACTIONS: Cascade Caverns, Cave-Without-A-Name, Wonder Cave, Inner Space, San Antonio Missions.

WONDER WORLD

Mailing address: P.O. Box 1369, San Marcos, TX 87666   *Phone:* (512) 392-671   *Directions:* Between Austin and San Antonio off Interstate 35 at San Marcos Exit.

This small cave, discovered while drilling for water in the hill section of central Texas, has little appeal to the connoisseur of show caves. It is developed along a fault line, has narrow steeply descending passageways, and few formations. Of geologic interest, however, is the display of fossils embedded in the limestone. During the course of the tour visitors descend 160 feet. An elevator at this point takes the visitor back to the surface; if he has bought a combination ticket, he continues to rise up another hundred feet to the top of a metal tower where he may view the Texas countryside from a glassed-in platform.

¶OPEN: all year summer 8:00 A.M. to 8:00 P.M.; winter 9:00 A.M. to 5:00 P.M. ¶GUIDED TOUR: 1½ hour   ¶ON PREMISES: snack bar, gift shop, Texas Observation Tower, Anti-gravity House, Texas Wildlife Park, picnicking   ¶NEARBY: restuarant, motel, camping, trailer camp   ¶NEARBY ATTRACTIONS: Inner Space, Natural Bridge Caverns, Cascade Caverns, Cave-Without-A-Name, Aquarena Springs.

UTAH

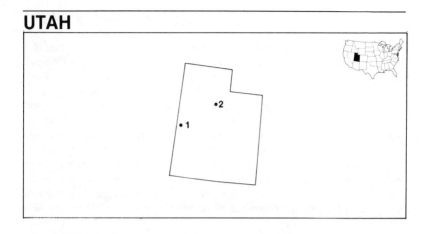

1 Crystal Ball Cave 2 Timpanogos Cave National Monument

THE COLORADO PLATEAU, a region of mountains in addition to table lands, includes part of Arizona, New Mexico, Colorado, and eastern Utah. In the western part of the state lie salt flats and an inland sea, Great Salt Lake. Limestone is found in many places in small deposits, but there is no great cave area in Utah. Many arches, natural bridges, and shelter caves are found in the sandstones of the Plateau. They were known, named, and occupied by Indians who built remarkable cliff dwellings in the shelters of the region. The locations and descriptions of these caves are excluded from this guide at the request of the Park Service, the ruins being extremely fragile and subject to damage by visitation.

CRYSTAL BALL CAVE

Mailing address: 405 Victoria Drive, Elko, NV 89801   *Phone:* (702) 738-7895
Directions: About 35 miles north of U.S. 6 and 50 on the road to Gandy. Reservations required, phone ahead.

Thomas Sims, a rancher searching for a lost sheep, found instead the entrance to a cave. The open sinkhole led to a corridor decorated with crystals and curious formations that resemble gigantic crystal-encrusted puff-balls. Although a generator has been installed to provide electric lighting, this cave is actually "wild," with tours offered by appointment only. Mr. Sims, the owner, is attempting to interest the National Park Service in the cave, which has unusual crystals. There are no signs leading to the cave so that it is impossible to locate without a guide; therefore it is essential that an appointment be made before attempting to visit.

¶OPEN: on request only. No admittance without permission.   ¶ON PREMISES: no facilities. This is a remote area that requires adequate gasoline; bring drinking water. Campsite at Baker (45 miles)   ¶NEARBY ATTRACTIONS: Lehman

214

Caves National Monument is about 50 miles to the south.

TIMPANOGOS CAVE NATIONAL MONUMENT

Mailing address: R.F.D. 2, Box 200, American Fork, UT 84003   *Phone:* (801) 756-4497   *Directions:* About 7 miles east of Interestate 15 on State Highway 80.

The Timpanogos Cave system consists of three small caves connected by man-made tunnels. The first, Hansen Cave, was discovered in 1887; the other two, Timpanogos Cave and Middle Cave, were discovered in 1921. To reach the entrance from Monument headquarters, follow the trail 1½ miles up the steep side of Mount Timpanogos. From the entrance, about 1000 feet above the canyon floor, are outstanding views of the Wasatch Mountains, Utah Valley, and American Fork Canyon. The cave is covered by a filigree of pink and white

Timpanogos Cave National Monument, Utah.

translucent crystals, with tiny pools of water reflecting its beauty. A huge heart-shaped stalactite, "The Great Heart of Timpanogos," is ringed with twisting helictites and aragonite crystals (see photo).

¶OPEN: May through October from 8:00 A.M. to 3:00 P.M. except Memorial Day to Labor Day when it is open to 6:00 P.M.   ¶GUIDED TOUR: 3 hours includes 3-mile round trip foot trail to cave entrance (only access).   ¶ON PREMISES: snack bar, gift shop, picnicking   ¶NEARBY: camping, motel, hotel, restaurant, trailer camp   ¶NEARBY ATTRACTIONS: Scenic Alpine Loop and the Mount Timpanogos Scenic Area, Bingham Copper Mine (world's largest open-cut mine), Utah Lake, Camp Floyd Historic State Park, Bridal Veil Falls, Rockport State Park and Reservoir.

VERMONT

SMALL SHELTER CAVES, marble caves, and boulder caves are found in Vermont, but no show caves.

VIRGINIA

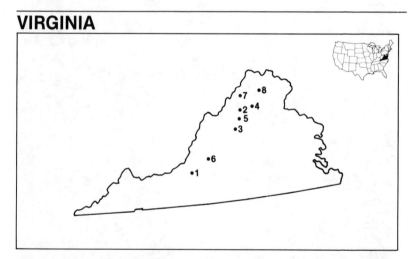

1 Caverns of Natural Bridge, Virginia 2 Dixie Caverns 3 Endless Caverns 4 Grand Caverns 5 Luray Caverns 6 Massanutten Caverns 7 Shenandoah Caverns 8 Skyline Caverns

VIRGINIA CONTAINS three distinct regions: in the east the sandy Atlantic Coastal Plain, the Piedmont Plateau in the center, and the Blue Ridge and Appalachian Mountains to the west where the caves are located. The cave-bearing limestones exposed in these mountains are from 400 to 450 million years old. The Shenandoah Valley that crosses the state diagonally between the mountain ridges not only provides a fertile corridor, but most of the caves are found here. Some, which have become world famous, have been a point of destination for travelers for more than a hundred years. Historically the caves were used by Indians for shelter and sanctuary. Early settlers mined saltpeter in the caves to make gunpowder during the War Between the States. The first show cave in the United States was Weyer's Cave, known today as Grand. Caverns. The success of this venture encouraged other entrepreneurs to open dozens of caves throughout the United States.

CAVERNS OF NATURAL BRIDGE, VIRGINIA

Mailing address: P.O. Box 147, Natural Bridge, VA 24578 *Phone:* (703) 291-2936 *Directions:* On U.S. 11 between Roanoke and Lexington at Natural Bridge.

Natural Bridge, Virginia, long known as an historic landmark, was made famous by artists in the 1700s. The main north-south trail through the Appalachian valley of Virginia crossed over the bridge and it attracted many travelers during the 19th century. The route over the bridge is utilized today by U.S. Highway 11. The bridge, remnant of a collapsed cave, spans a narrow fissure 215 feet above Cedar Creek. There are several caves on the property including Saltpeter Cave, Doll House Cave, and on the hill above the bridge, a cave known as Buck Hill Cave. In 1977 a tunnel was dug into Buck Hill Cave, and lights, trails, and steps were installed. It was renamed Caverns of Natural Bridge, Virginia. There is a flowing stream in the cave, a waterfall, and rimstone dams and pools. The cave is active and glistening as the formations continue to grow in the humid atmosphere.

¶OPEN: all year 10:00 A.M. to 6:15 P.M. in winter; 9:00 A.M. to 7:15 P.M. in summer ¶GUIDED TOUR: 45 minutes to 1 hour ¶ON PREMISES: gift shop, picnicking ¶NEARBY: all facilities· ¶NEARBY ATTRACTIONS: Dixie Caverns, Blue Ridge Parkway, Booker T. Washington National Monument

DIXIE CAVERNS

Mailing address: 5596 Highfields Road SW, Roanoke, VA 24000 *Phone:* (703) 389-7588 *Directions:* Off Interstate 81 about 5 miles west of Salem.

Dixie Caverns on the northeast slope of Cave Hill near Fort Lewis Mountain was opened to the public in 1922. Only 200 yards from Lee Highway (Route 11), the north-south corridor through western Virginia, it has enjoyed continuing popularity. The developed section, about ¼-mile long, consists of narrow passages and large chambers on three levels, with several intersecting and cross

Dixie Caverns, Virginia.

channels. The rooms feature many formations, including the "Wedding Bell" shown in the photograph above. This is also the home of the "Dixie Salamander," a salamander first discovered here and named after the cave. This protected species is shy, avoids the light, and is rarely seen on a tour.

¶OPEN: daily all year from 8:00 A.M. to 4:30 P.M. except summer when open until 9:00 P.M.   ¶GUIDED TOUR: 40 minutes   ¶ON PREMISES: gift shop, camping, trailer camp, pottery shop, picnicking ¶NEARBY: all facilities ¶NEARBY ATTRACTIONS: Blue Ridge Parkway, Organ Cave (West Virginia), Lost World (West Virginia), Claytor Lake State Park, Booker T. Washington National Monument.

ENDLESS CAVERNS

Mailing address: New Market, VA 22844   *Phone:* (703) 740-8877   *Directions:* About 2 miles off Interstate 81 at New Market on U.S. 11.

When two boys chased a rabbit down a hole on the slope of Massanutten Ridge in 1879, they accidently discovered the entrance to Endless Caverns. The owner, Ruben Zirkle, inspired by the success of Luray Caverns 11 miles to the

Endless Caverns, Virginia.

east, built trails and a dance floor to attract visitors. Twisting intersecting passageways and a profusion of formations that nearly block the trail create the illusion that the cave is endless. In 1920 new lights, trails, and walkways were installed. Mountaineers from The Explorers Club of New York attempted to discover the "end" of Endless Caverns. At the limit of their search, as reported in the national press, they left a bottle with the names of team members. The bottle has been advanced several times by exploring parties but none claim to have come to the end of the cave. The Shenandoah Valley of Virginia has some of the finest show caves in the United States, and Endless Caverns is one of the best of them.

¶OPEN: daily all year 9:00 A.M. to 5:00 P.M. ¶ON PREMISES: snack bar, gift shop, camping, trailer camp, picnicking ¶NEARBY: all facilities ¶NEARBY ATTRACTIONS: Luray Caverns, Shenandoah Caverns, New Market Battlefield, Massanutten Caverns, Grand Caverns, Skyline Drive, Natural Chimneys, Shenandoah National Park.

GRAND CAVERNS

Mailing address: P.O. Box 478, Grottoes, VA 24441 *Phone:* (703) 249-2451
Directions: About ½-mile west of U.S. 340, 16 miles north of Waynesboro in the town of Grottoes.

Grand Caverns, Virginia

Overlooking the South Fork of the Shenandoah River is Grand Caverns, originally called Weyer's Cave after its discover, Bernard Weyer. He opened the cave for visitors in 1806, which makes it the oldest show cave in the United States. Early visitors to the cave include Thomas Jefferson and the noted artist Porte Crayon, whose scenes of the cave spread its fame. A visitor during the War Between the States was Stonewall Jackson, who quartered his troops nearby. When Fountain Cave was discovered a few hundred yards from Weyer's Cave they were both exhibited under the name "Grottoes of the Shenandoah." Later Fountain Cave was closed and Weyer's Cave became known by its present name, Grand Caverns. The disk-shaped calcite formations called "shields" are rare in caves but Grand Caverns has a remarkable display of them. Outstanding is the "Giant Oyster Shell" formation. Grand Caverns is now a state park operated by the state of Virginia.

¶OPEN: daily all year from 9:00 A.M. to 5:00 P.M.; extended hours between Memorial Day and Labor Day   ¶GUIDED TOUR: 1 hour   ¶ON PREMISES: snack bar, gift shop, picnicking, recreation area   ¶NEARBY: camping, trailer camp, restaurant, motel, hotel   ¶NEARBY ATTRACTIONS: Natural Chimneys, Endless Caverns, Luray Caverns, Massanutten Caverns, Natural Bridge Caverns, Shenandoah Caverns, Woodrow Wilson Birthplace, Big Meadows Recreational Area, Home of Thomas Jefferson, Home of James Monroe, North River Recreational Area, Hone Quarry Recreational Area.

LURAY CAVERNS

Mailing address: P.O. Box 748, Luray, VA 22835   *Phone:* (703) 743-6551
Directions: On U.S. 211 Bypass about 1 mile west of Luray.

Luray Caverns was discovered in 1878 by Benton P. Stebbins and Andrew J. Campbell, who were searching for a cave to exhibit that would be as successful as Weyer's Cave in Augusta County. The cave, about a mile from Luray, was exhibited within a few days of the discovery and has attracted visitors in such numbers that today it is the most popular cave on the eastern seaboard. One of the finest show caves in the United States, it has spectacular views and formations at every point. The complex series of passages is partly blocked by the massive formations that create walls and partitions within the large rooms. In addition to the natural features of the cave is the underground Stalacpipe Organ, which has special hammers that gently strike individual stalactites, causing ethereal harmonies to reverberate through the cavern. Included in the price of admission is the Car and Carriage Caravan, an exhibition of 75 authentically restored cars, carriages, and costumes, and Luray Singing Tower, a carillon of 47 bells.

¶OPEN: daily all year from 9:00 A.M. to 6:00 P.M. except summer to 7:00 P.M. and winter to 4:00 P.M.   ¶GUIDED TOUR: 1 hour   ¶ON PREMISES: restaurant,

Luray Caverns, Virginia.

snack bar, gift shop, 2 motels, golf course, tennis, picnicking ¶NEARBY: camping, trailer camp ¶NEARBY ATTRACTIONS: Endless Caverns, Shenandoah Caverns, New Market Battlefield, Massanutten Caverns, Grand Caverns, Skyline Caverns, Skyline Drive, Natural Chimneys, Shenandoah National Park.

MASSANUTTEN CAVERNS

Mailing address: Keezletown, VA 22832 *Phone:* (703) 269-4821 *Directions:* Between Elkton and Harrisonburg turn north on State Highway 620 about 3 miles to Keezletown.

The Massanutten Mountain divides the northern end of the Shenandoah Valley, rising like a broaching whale out of the rolling plain. On the southwest end of this mountain lies the entrance to Massanutten Caverns, overlooking Cub Run, a tributary of the South Fork of the Shenandoah River. Discovered in 1892, the cave was not opened to the public until 1926. A quarter-mile passageway, nearly level, winds through some of the finest cave scenery in the Shenandoah Valley. Fragile, delicate formations decorate the walls along the trail and each room, providing variety to the scene.

Massanutten Caverns, Virginia.

Two levels are known to exist below the main tour level, but they are not exhibited. The passages on the tour level are constricted, but easy to traverse. The cave is small but beautiful. It consists of a connected series of rather narrow short and long corridors and numerous alcoves. The highest ceiling is not more than 25 feet above the floor. The Ball Room is the largest room in the caverns, with a length of about 100 feet. There is no visitors' building at the cave. A shelter provides a place to wait for the next tour. During the winter season it is advisable to call ahead for a reservation.

¶OPEN: all year June 1 to Labor Day daily 10:00 A.M. to 6:00 P.M. Spring and fall daily 11:00 A.M. to 5:00 P.M. Winter schedule weekends only 12:00 noon to 5:00 P.M.   ¶GUIDED TOUR: 1 hour   ¶ON PREMISES: picnicking   ¶NEARBY: all facilities in Harrisonburg   ¶NEARBY ATTRACTIONS: Endless Caverns, Luray Caverns, Shenandoah Caverns, Grand Caverns, Natural Bridge Caverns, New Market Battlefield, Skyline Drive, Natural Chimneys, Shenandoah National Park.

SHENANDOAH CAVERNS

Mailing address: Box 1, Caverns Road, Shenandoah Caverns, VA 22847
Phone: (703) 477-3115   *Directions:* Off Interstate 81 at Shenandoah Caverns Exit 68 about 4 miles north of New Market.

Shenandoah Caverns, Virginia.

Construction of the Southern Railroad through the Shenandoah Valley exposed Shenandoah Caverns on Cave Hill in 1884. Previously there were several pits and small caves known in the hill west of the Shenandoah River, but the excavations by the railroad crews revealed the major cavern of the region. In 1922 the cave was opened to visitors following installation of an elevator, trails, lights and stairs. The elevator made the tour one of the easiest and most popular in the valley. Long spacious corridors reveal dramatic scenes of formations that in places completely fill the passage. The cave well deserves the name Shenandoah as representative of the remarkable caverns that have made this valley famous. The main cave is largely on one level with slight changes in grade. There are two other levels not shown on the tour. Legend has it that the cave was known and visited by Indians, but it is doubtful that anybody entered the cave prior to its discovery by railroad workers.

¶OPEN: all year daily from 9:00 A.M. to 4:30 P.M. except in summer when hours are extended to 6:30 P.M.   ¶GUIDED TOUR: 1 hour   ¶ON PREMISES: snack bar, gift shop, picnicking, elevator in cave   ¶NEARBY: restaurant, motel, camping, trailer camp, hotel, cabins   ¶NEARBY ATTRACTIONS: New Market Battlefield, Endless Caverns, Luray Caverns, Massanutten Caverns, Grand Caverns, Skyline Caverns, Skyline Drive, Natural Chimneys.

SKYLINE CAVERNS

Mailing address: P.O. Box 193, Front Royal, VA 22630   *Phone:* (703) 635-4545   *Directions:* On U.S. 340 about 1 mile south of Front Royal at the entrance to the Skyline Drive.

Skyline Caverns, Virginia.

Skyline Drive in the Blue Ridge Mountains attracts hundreds of thousands of visitors each year. In 1937, geologist Walter S. Amos deduced from exposure of the limestone plus the evidence of a shallow cave overlooking the Shenandoah River that potential for a large cave existed in the ridge. After several trial excavations, he was rewarded by the discovery of a hole leading into the ridge only a few hundred yards from the scenic Drive. His first view of the cave revealed only barren waterworn passages devoid of formations. Undaunted, he explored further, discovering not only several large rooms but also the most remarkable collection of rare anthodite flowers now exhibited in this country. The passage containing these rock formations was filled with clay and required careful hand work to clear a trench for viewing by visitors. The anthodites are extremely fragile with hairlike crystals radiating from a central stalactite. Their stark whiteness is in sharp contrast to the smooth reddish walls. The tour of the cavern also includes a view of an underground stream and waterfall that plunges 37 feet to lower undeveloped levels.

¶OPEN: all year daily 9:00 A.M. to 5:00 P.M. Extended hours in summer months. ¶GUIDED TOUR: 1 hour   ¶ON PREMISES: restaurant, snack bar, gift shop, picnicking   ¶NEARBY: motels, camping, trailer camp   ¶NEARBY ATTRACTIONS: Luray Caverns, Endless Caverns, Shenandoah Caverns, New Market Battlefield, Skyline Drive, Shenandoah National Park.

WASHINGTON

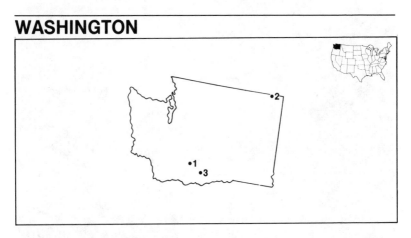

1 Ape Cave 2 Gardner Cave 3 Ice Cave

THE STATE OF WASHINGTON has a varied landscape that includes the Pacific Coastal Ranges in the west, the Columbia Plateau in the south central section, and the Rocky Mountains in the northeast. Volcanic activity has poured out great fields of lava, creating cinder cones, volcanic peaks, and extensive lava tubes in the Columbia Plateau. A few limestone caves are scattered about the state in small pockets of limestone, but none are large.

APE CAVE

Mailing address: Gifford Pinchot National Forest, 500 West 12th Street, Vancouver, WA 98660 *Phone:* (206) 696-4041 or (1-206) 238-5244 (Ranger Station) *Directions:* About 45 miles off Interstate 5 on State Highway 504 to Spirit Lake.

Several thousand years ago molten viscous pahoehoe lava flowed down the slopes of Mount St. Helens in what is now Pinchot National Forest. As the surface lava cooled and solidified, the inner molten core continued to flow underneath. Eventually it drained away, leaving intact a hollow lava tube two miles long, the longest known in the United States. In 1951, a bulldozer operator clearing brush nearly dropped into the lower entrance, a 20-foot-deep hole 15 feet around. A local group of outdoorsmen, the St. Helens Apes, explored the cave and named it Ape Cave for their organization. While there are several distinct levels within the cave, it is basically one continuous tunnel with all of the features characteristic of lava tubes visible in the more than 11,000 feet that have been explored. There are two entrances formed by the collapse of the thin roof, and light can be seen in several places where breakdown domes approach the surface. The main entrance is about 100 yards from the parking area. There are no guides and it is necessary to bring adequate lights. A map is available at the ranger station. It is recommended that a sweater be worn as the cave has an average temperature of 42° F. during the summer.

¶OPEN: Memorial Day through November 15 weather permitting. Daylight hours ¶SELF-GUIDED TOUR: Map available from Spirit Lake Ranger Station. Bring lights and wear sturdy shoes. Trail marked to cave ¶ON PREMISES: camping, trailer camp, picnicking ¶NEARBY: scattered facilities ¶NEARBY ATTRACTIONS: Mount St. Helens Winter Sports Area, Spirit Lake.

GARDNER CAVE

Mailing address: Crawford State Park, Metaline, WA 99152 *Phone:* (509) 456-4169 *Directions:* Take a dirt road at the northern limit of Metaline off State Highway 6 about 10 miles to parking area of State Park.

The entrance to Gardner Cave is a collapsed sinkhole in a densely forested area northwest of Z Canyon. The cave, a feature of Crawford State Park, was partially developed as a tourist attraction in 1959. Only about a half-mile from the Canadian border, it is probably the northernmost limestone cave in the contiguous United States. A relatively small cave about 1000 feet long, it has suffered considerable vandalism since its discovery in 1903 by homesteader Ed Gardner. In 1921 the cave and surrounding area were deeded to the state. Access is by wooden ladder to the floor of the cave. The corridors are easily traversed. Several streams intersect the passages; rimstone pools, moonmilk, and large dripstone formations are found in the cave.

¶OPEN: Memorial Day through October 15 during daylight hours   ¶SELF-GUIDED TOUR: bring flashlights and sturdy shoes   ¶ON PREMISES: camping, picnicking   ¶NEARBY: all facilities in Metaline (10 miles)   ¶NEARBY ATTRACTIONS: Sullivan Lake and Dam, St. Paul's Mission Recreation Area, Coulee Dam National Recreation Area.

ICE CAVE

Mailing address: Gifford Pinchot National Forest, 500 West 12th Street, Vancouver, WA 98660   *Phone:* (206) 696-4041 or (1-206) 238-5244   *Directions:* From Hood about 25 miles north on State Highway 141 to Trout Lake. West 5 miles to Ice Cave Forest Campground.

This well-known lava tube on the slopes of Mount Adams was used as an ice supply for the towns of Hood River and The Dalles in early pioneer days. There are several entrances where the ceiling has collapsed. Access by ladder leads to an ice pool that gives the cave its name. Located on Forest Service land within the Ice Cave Forest Campground, this is a true *glacière* (perpetual ice cave), where cold, freezing air trapped during the winter forms ice that never melts completely during the summer. There is only limited time during the year when the cave is accessible, as it is snowed in until mid-June and refreezes in October or November.

¶OPEN: Memorial Day through November 15 weather permitting. Daylight hours   ¶SELF-GUIDED TOUR: Map available from Trout Lake Ranger Station. Bring lights and wear sturdy shoes. Trail marked to cave   ¶ON PREMISES: camping, trailer camp, picnicking   ¶NEARBY: scattered facilities   ¶NEARBY ATTRACTIONS: Government Mineral Springs, Columbia River resort area.

WEST VIRGINIA

1 Lost World Caverns 2 Organ Cave 3 Seneca Caverns 4 Smoke Hole Cavern

THE MOUNTAINOUS EASTERN AND southern half of West Virginia contains the folded and faulted rocks of the Allegheny Mountains, which have accumulated sediments to a thickness of 40,000 feet. The hilly Appalachian Plateau to the west has rocks that are more flat-lying. In addition to huge reserves of coal, West Virginia has large amounts of limestone rock and thousands of caves. The caves are extensive and rival any in the county, but the difficulty of access has limited their popularity and commercial development. Historically the caves are rich in folklore and tales of treasure; many were sources of saltpeter for gunpowder by early settlers. Indian artifacts have been found in some of the caves.

LOST WORLD CAVERNS

Mailing address: P.O. Box 247, Lewisburg, WV 24901   *Phone:* (304) 645-1658
Directions: About 2 miles north of Interstate 64 on U.S. 219.

On the east flank of Weaver Knob a vertical shaft leads 115 feet down to one of the finest cave rooms in West Virginia. Discovered in 1942 by speleologists using ropes and ladders, it was known as Grapevine Cave until 1970 when a horizontal tunnel was dug into the main chamber. Renamed Lost World Caverns, it now features an easy walking tour and is noted for large impressive formations. On the floor is an uneven mass of breakdown but the ceiling is nearly flat. A trail encircling the room provides a close-up view of the formations that were so highly prized by the discoverers when they entered the room through the ceiling entrance.

¶OPEN: all year 9:00 A.M. to 5:00 P.M. except extended hours in summer to 7:00 P.M.   ¶SELF-GUIDED TOUR: taped narration at vantage points in the cave. About 1 hour   ¶ON PREMISES: snack bar, gift shop, picnicking   ¶NEARBY: all facilities ¶NEARBY ATTRACTIONS: Organ Cave, Greenbrier Resort, Hot Springs, Seneca Caverns, Smoke Hole Cavern, White Sulphur Springs, Old Rehobeth Church,

Lost World Caverns, West Virginia.

Droop Mountain Battlefield, Watoga State Park, Donthan State Park, Dixie Caverns (Virginia).

ORGAN CAVE

Mailing address: Route 3, Box 381, Ronceverte, West Virginia 24970 *Phone:* (304) 647-5551 *Directions:* Off U.S. 219 about 3 miles south of Ronceverte.

Organ Cave, located at the head of a peaceful green valley, was one of the confederate Army's military secrets during the War Between the States more than a hundred years ago, as it was a site for the manufacture of saltpeter for gunpowder. Today the guided tour takes visitors back to the time when conscripted men and boys crawled through narrow passages, scooped up the dirt from the floor, and deposited it in saltpeter leaching vats deep within the cave. These vats, 37 in all, are in remarkable condition. The wood is well preserved and the workmanship excellent. Although the cave is not lighted well enough to bring out the beauty of the formations or the convolutions of the natural walls, a visit is well worth while to see what is probably the best preserved collection of saltpeter workings in the country.

¶OPEN: Memorial Day through October 9:30 A.M. to 6:00 P.M. ¶GUIDED TOUR: 1 hour ¶ON PREMISES: gift shop, camping, picnicking ¶NEARBY: camping, motel, trailer park, restaurant, cabins ¶NEARBY ATTRACTIONS: Lost World, Stone Church, White Sulphur Springs, Blue Bend recreation area.

SENECA CAVERNS

Mailing address: Riverton, WV 26814 *Phone:* (304) 567-2691 *Directions:* Off U.S. 33 at Riverton about 3 miles east.

The entrance to Seneca Caverns is a shallow sink that has been known for hundreds of years. The cave supposedly was a refuge for the Seneca Indians, and a legend tells of Chief Bald Eagle living for a short time within the cave. In 1930 the introduction of lights and trails opened the cave well beyond the deepest penetration of the Indians. A horizontal, sinuous passage leads under the ridge where sparkling white formations decorate the walls.

¶OPEN: April 1 through October 31 daily 8:00 A.M. to 7:00 P.M. ¶GUIDED TOUR: 45 minutes ¶ON PREMISES: snack bar, gift shop, camping, trailer camp, picnicking ¶NEARBY ATTRACTIONS: Smoke Hole Cavern, Sinks of Gandy, Seneca Rocks, Blackwater Falls State Park, Spruce Knob (highest point in West Virginia), Endless Caverns (Virginia), Shenandoah Caverns (Virginia).

SMOKE HOLE CAVERN

Mailing address: Moorefield, WV 26836   *Phone:* no phone   *Directions:* On
State Highway 4 and 28 about 8 miles west of Petersburg.

Smoke Hole Cavern has a long history dating from Indian times when the cave
was used as a place to smoke meat. The entrance, overlooking Jordan Run,

Smoke Hole Cavern, West Virginia.

was a favored Indian campsite. During the Civil War it served as a storage area for ammunition. Later the isolated valley and cave became a secret spot to make corn whiskey. A still and jugs for bottling the brew are displayed in the cave. The tour follows a small stream along the lower level and then ascends a stairway to the upper main room of the cave. There the ceiling bristles with thousands of tiny soda straw stalactites. The cave is noted for the ribbon stalactites that drape the walls plus the evidence of its long history of occupation.

¶OPEN: April 1 through December 1 daily 8:00 A.M. to 7:00 P.M.   ¶GUIDED TOUR: 50 minutes   ¶ON PREMISES: restaurant, snack bar, gift shop   ¶NEAR-BY: all facilities   ¶NEARBY ATTRACTIONS: Champ Rocks, Seneca Rocks, Blackwater Falls State Park, Lost River State Park, Seneca Caverns, Shenandoah Caverns (Virginia), Endless Caverns (Virginia).

WISCONSIN

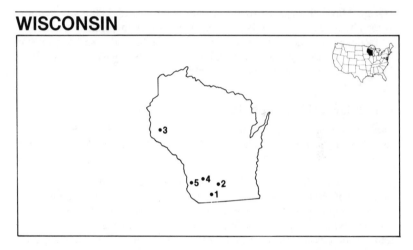

1 Badger Mine (artificial cave) 2 Cave of the Mounds 3 Crystal Cave 4 Eagle Cave 5 Kickapoo Indian Caverns

WISCONSIN IS DIVIDED north-south into two sections: the Great Plains to the south, and the Superior Highlands to the north. All of the caves are found in the southern part of the state where dolomite, a form of limestone, was deposited about 400 million years ago. Although most of the state is covered by glacial deposits, the Driftless Section (at the intersection of Minnesota, Iowa, and Wisconsin) was never glaciated. Therefore fragile landforms such as natural arches, buttes, and rock towers survive on the surface, and caves and underground streams below. Some of the caves were visited by Indians. Others were discovered by prospectors in search of lead and zinc.

BADGER MINE (artificial cave)

Mailing address: Shullsburg, WI 53586   *Phone:* (608) 965-4860   *Directions:* On State Highway 11 in the town of Shullsburg.

While Badger Mine, which lies within the city limits of Shullsburg, is not a cave, it has many of the attractions of a cave. It is operated by the La Fayette Association for Retarded Children, Inc., for the benefit of unfortunate youngsters. Actually it is a lead mine, first operated in 1827, that extends for miles beneath the city. The tourist route covers the main passage for half a mile, and from it side-tunnels and workings can be seen. The original miners in the area were Winnebago Indians who mined for lead in the very early 1800s. There is also a park and museum operated in connection with the mine. It was opened to the public in 1965.

¶OPEN: May 1 through November 1 daily 9:00 A.M. to 5:00 P.M. ¶GUIDED TOUR: about 30 minutes ¶NEARBY: snack bar, gift shop, picnicking   NEARBY: all facilities ¶NEARBY ATTRACTIONS: Cave of the Mounds, New Glarus State Park, Nelson Dewey State Park, Eagle Cave, Governor Dodge State Park, Yellowstone Lake, General Grant's home.

CAVE OF THE MOUNDS

Mailing address: Brigham Farm, Blue Mounds, WI 53517   *Phone:* (608) 437-3355   *Directions:* ¼-mile off U.S. 18 and 151 about 25 miles west of Madison.

The discovery of Cave of the Mounds in 1939 took place in an instant with the explosion of 1600 pounds of dynamite in a quarry east of Blue Mounds. There had been no prior clue that a cave lay concealed behind the cliff face. The blast exposed about 40 feet of the main passage, dividing the cave into two sections. Exploration revealed the cave as we see it today, without further damage. The cave, opened to the public in 1940, has a map that looks like a giant inverted fish hook with the entrance at the side of the stem. The South Cave is a nearly straight corridor with 30-foot-high ceilings and an impressive row of stalagmites on the floor. The tour proceeds to the end, and returns to the entrance where a stairway leads to the North Cave and the curved part of the "fish hook." This section has many formations, including helictites, lily pads, soda straws, and cave pearls. Cave of the Mounds is a favorite area for classroom tours by local schools, and provides a unique educational experience for young people.

¶OPEN: April 1 through October 31 daily 9:00 A.M. to 5:00 P.M. except Memorial Day to Labor Day when hours are extended to 7:00 P.M. November, weekends only ¶GUIDED TOUR: 45 minutes to 1 hour ¶ON PREMISES:

Cave of the Mounds, Wisconsin.

restaurant, snack bar, gift shop, picnicking ¶NEARBY: motel, camping, trailer camp, hotel, cabins ¶NEARBY ATTRACTIONS: Little Norway, Mount Horeb, Mineral Point, Eagle Cave, Badger Mine, House on the Rock, Blue Mounds State Park.

CRYSTAL CAVE

Mailing address: Spring Valley, WI 54767   *Phone:* (715) 778-4414   *Directions:* 1 mile west of Spring Valley on State Highway 29.

Crystal Cave, Wisconsin.

In 1881 a young farm boy was attracted to a deep pit in the wooded foothills of the Eau Galle River and explored it with rope and candles. Although he told his friends of the great cave that lay beneath the hill, it remained a local curiosity and was only partially explored until 1942 when Henry Freide developed it and named it Crystal Cave. The guided tour descends three distinct levels through a series of rooms that have been excavated to provide adequate headroom. Included on the tour are examples of onyx formations, cave pearls, and gypsum flowers.

¶OPEN: daily all year from 9:00 A.M. to 5:00 P.M.; extended hours in spring and fall to 7:00 P.M.; in summer until 8:00 P.M. ¶GUIDED TOUR: 1½ hours ¶ON

PREMISES: restaurant, snack bar, gift shop, camping, picnicking, fishing and swimming ¶NEARBY: motel, cabins, trailer camp ¶NEARBY ATTRACTIONS: Spring Valley Dam, Cedar Lake, Lake Pepin, Lake Tainter.

EAGLE CAVE

Mailing address: Route 2, Blue River, WI 53518   *Phone:* (608) 537-2446   *Directions:* Off State Highway 60 about 15 miles south of Richland Center.

In 1849, while hunting along the Wisconsin River, Peter Kinder wounded a bear. He followed it into the entrance of a cave, discovering a labyrinth of passageways now known as Eagle Cave. The cave achieved local popularity but history fails to tell us what happened to the bear. In 1936 the cave was opened to the public; the surrounding land has been set aside as a park for summer and winter activities including swimming, hiking, boating, skating and sledding. Regular tours through the cave visit many features in remote rooms. On Saturday nights during the summer, square dances are held in Black Hawk's Ballroom. This chamber, 250 feet long and 50 feet wide, has an artificial entrance from the outside as well as access to the main cave. During the summer, picnicking inside the cave is popular on hot and rainy days.

A scout program has been attended by more than 100,000 young people over the years. It includes a spelunking tour to the undeveloped portions of the cave and overnight camping underground, where the temperature is a constant 53° F. Other facilities are available outdoors for the scouts, making this a popular program for many troops. Reservations should be made well in advance for groups that wish to participate in this cave-camping experience.

¶OPEN: daily all year 9:00 A.M. to 8:00 P.M.   ¶GUIDED TOUR: 1 hour   ¶ON PREMISES: snack bar, gift shop, camping, trailer camp, picnicking, square dancing in cave ¶NEARBY: motel, hotel, restaurants ¶NEARBY ATTRACTIONS: House on the Rock, Wisconsin Dells, Tower Hill State Park, Kickapoo Caverns, Cave of the Mounds, Badger Mine.

KICKAPOO INDIAN CAVERNS

Mailing address: Route 1, Box 222, Wauzeka, WI 53826   *Phone:* (608) 875-5223   *Directions:* On State Highway 60 about 10 miles east of Prairie Du Chien.

The entrance to Kickapoo Indian Caverns overlooks the Little Kickapoo River Valley near the town of Wauzeka. Known and visited by local Indians, it was first reported in the early 1800s by prospecting lead miners. The word "kickapoo" comes from the Algonquin Indian word meaning "moving about" or "flitting." True to their name, many members of this tribe migrated to the southwest and Mexico.

Entrance to the cave is through a keyhole-like opening to the rear of the Visitors' Building built in 1947. Excavation at that time revealed Indian artifacts and abandoned miners' equipment. Most of these relics are displayed in a small museum but some are still in place in the cave where they were discovered. The tour descends a fissure-like corridor with a level ceiling and sloping floor to where several passages branch off and lead to a series of well-decorated rooms.

¶OPEN: May through October daily 9:00 A.M. to 5:00 P.M.; Memorial Day to Labor Day 8:00 A.M. to 6:00 P.M. ¶GUIDED TOUR: 40 minutes ¶ON PREMISES: trading post and Indian museum, picnicking ¶NEARBY: restaurant, camping, trailer camp, motels, hotel ¶NEARBY ATTRACTIONS: Phetteplace Museum, Villa Louis, Spook Cave (Iowa), Effigy Mounds, Wyalusing State Park, Pikes Peak State Park, Nelson Dewey State Park, Old Fort Crawford.

WYOMING

WYOMING IS DIVIDED by the Rocky Mountains, which run north-south through the state. The Wyoming Basin in the center of the state separates the Middle Rocky Mountains from the Southern Ranges. There are many caves within the state, but none of them are now open to the public. Several tufa caves formed by the minerals from hot springs are found in Wyoming.

GLOSSARY OF CAVE TERMS

THE FOLLOWING GLOSSARY covers words used and referred to in the text and is not intended to be a complete listing of speleological terms.

ALABASTER: A form of gypsum that has been widely used for carving statuary.

ANTHODITE: A flower-like formation, usually in aragonite or calcite, forming a spray of needles from a center stalk.

ARAGONITE: A form of calcium carbonate sometimes found in caves, usually in the form of needle-like crystals.

ARTIFACTS: Articles made by man and discovered at a later date. They are usually evidence of man's previous occupancy.

BACON RIND: A thin drapery that generally forms on sloping walls; it has bands of different colors. When this formation is lighted from behind, it resembles a strip of bacon.

BEDDING PLANES: Horizontal boundaries between two adjacent layers of sedimentary rock.

BLIND CAVE FISH: Rare fish that are found in deep caves in the Ozarks and southern portions of the United States generally.

BOXWORK: Thin veins of calcite projecting from the walls and ceilings of some caves. Usually they are found in roughly rectangular patterns.

BREAKDOWN: Found in a room or passage, these irregular blocks of stone have long ago collapsed from the ceiling.

BROOMSTICKS: Tall, thin stalagmites, usually only a few inches in diameter.

CALCITE: The most common mineral found in caves, it is composed of calcium carbonate, and often takes a crystalline form, which affects the growth of the various stalactites and stalagmites found in caves.

CALCIUM CARBONATE: This mineral, $CaCO_4$, occurs in nature as calcite and aragonite, and is the principal constituent of limestone.

CAVING: The sport of cave exploration. Synonymous with SPELUNKING.

COLUMN: A stalactite and stalagmite that have grown together to form a pillar that joins the floor and ceiling.

COMMERCIAL CAVES: Privately owned caves open to the public and charging admission.

CONGLOMERATE: A sedimentary rock composed mainly of cemented gravel.

CRUSTATION: Usually a thin covering of calcite of gypsum on cave walls.

CURTAIN: A thin and sometimes translucent hanging formation resembling draperies.

DOGTOOTH SPAR: Pure crystals of calcium carbonate, usually pointed, which are formed under water or in sealed chambers within the cave.

DOME PIT: A vertical enlargement of a cave corridor formed by solution, and not by breakdown, with a dome above and a pit below.

DRIPSTONE: A term used interchangeably with "flowstone" and sometimes with "stalagmite." *See* FLOWSTONE.

239

FLOWSTONE: Calcite that has coated the wall or floor of a cave.

FORMATION: Here broadly used to describe any secondary growth formed in a cave, and usually composed of calcite. The term can be applied to all types of growths that appear in caves. Synonymous with SPELEOTHEM.

FOSSILS: Remains of ancient animals or vegetable matter found in the bedrock. These may be either the actual remains of the animal or a cast formed by minerals.

GEODE: A hollow nodule of stone whose cavity is lined with crystals.

HELICTITES: Twisted or erratic formations, usually of calcite, which seem to defy gravity.

KARST: A limestone region marked by sinks and caverns. Usually there are no surface streams, and the limestone has been exposed by erosion.

LAVA TUBES: Tunnels left by the draining off of molten lava in level volcanic regions.

LILY PADS: Thin, horizontal calcite formations formed at the surface of still pools.

LIMESTONE: A rock of calcium carbonate. Of sedimentary origin, it is the most common stone in the formation of caves.

MATRIX: The stone that surrounds the cave: its floor, walls, and ceiling.

MOON MILK: Hydromagnesite, a mineral sometimes found in caves. It may take the appearance of cottage cheese or of a crystalline "snow."

NODULE: A small, generally rounded body usually somewhat harder than the enclosing sediment.

PALETTES: *See* SHIELDS.

PECCARY: A piglike animal the remains of which are sometimes found in caves.

PENDANT: A projection of stone hanging from the walls or ceiling of a cave. It is not the result of deposition.

PETROGLYPHS: Incised drawings or diagrams made by early man on rocks and ledges.

RIMSTONE POOLS: Pools formed by the deposit of calcite in slow-flowing streams.

ROCK SHELTERS: Shallow overhangs in a cliff face that were used by early man as a temporary or permanent shelter. Popularly called caves, they usually do not penetrate into the earth beyond the reach of light.

SALTPETER MINING: An industry developed to supply nitrate for the manufacture of gunpowder. Saltpeter (or "saltpeter earth") was dug in the caves, leached in vats, and then distilled to crystals.

SELENITE: Gypsum in the shape of needles.

SHIELDS: Disk-shaped formations that project from the walls and ceilings of some caves.

SHOW CAVES: Privately or publicly owned caves open to the public.

SODA STRAWS: Thin stalactites just the size of the drop that has formed the "straw."

SPELEOLOGIST: A specialist in the systematic study or exploration of caves.

SPELEOTHEM: A term applied to any secondary growth formed in a cave. *See* FORMATION.

SPELUNKER: A person who explores caves as a hobby or sport.

STALACTITE: Formation of calcium carbonate that hangs from the ceiling of a cave room or passage.

STALAGMITE: The companion to the stalactite formed by dripping water, usually from the stalactite above. This formation is more rounded than the stalactite, and is found on the floors and wall shelves of cave rooms.

TRAVERTINE: Calcium carbonate deposits that are alternately called "flowstone," "dripstone," or "cave onyx."

WATER TABLE: The level at which water is maintained by natural or artificial drainage from a cave.

INDEX

Names preceded by an asterisk (*) and *without a page number* are caves that are believed to be closed. Names preceded by an asterisk (*) and *with a page number* are names of caves not presently in use. The proper cave name will be found on the page indicated. Page numbers in italics refer to illustrations. The index does not include names listed in the cave descriptions under "Nearby Attractions." Names in quotations are the names given to formations or scenes in caves.